HOME ELECTRICAL REPAIRS

by

L.R.WAKELIN Grad. I.T.E.

EP Publishing Limited
1977

The *Invest in Living* Series

All About Herbs
Basic Woodworking
Edible Fungi
Getting the Best from Fish
Getting the Best from Meat
Home-Baked Breads and Scones
Home Decorating
Home Goat Keeping
Home Honey Production
Home-Made Butter, Cheese and Yoghurt
Home Maintenance and Outdoor Repairs
Home Poultry Keeping
Home Rabbit Keeping
Home Vegetable Production
Meat Preserving at Home
101 Wild Plants for the Kitchen
Pickles and Chutneys
Wild Fruits and Nuts

ISBN 0 7158 0491 X

Published 1977 by EP Publishing Ltd, East Ardsley, Wakefield, West Yorkshire WF3 2JN

Printed and bound in Great Britain by
Butler and Tanner Ltd, Frome and London

Contents

A Simple Guide to Understanding Electricity

Voltage is the pressure of force with which the current of electricity flows. The current amperes, usually referred to as amps, is the quantity of electricity passing through a circuit.

The voltage in a circuit can be looked at diagrammatically as a water tank, this provides the pressure or force. The amp can be likened to the amount of water which flows from a tank, i.e. the bigger the outlet the bigger the flow of water. Therefore the greater the power rating of an appliance the greater the amount of current consumed.

The watt is the amount of power consumed by the circuit. A 1 kw electric fire which is run for one hour will use 1 kw hour of electricity, and this is equal to one unit. The Electricity Board charges each domestic consumer for the amount of units used and, at present, this stands at 2·23p per unit.

The watt or kilo-watt (1000 watts) is the result of the pressure (volts) multiplied by the amount of current (amps). By transposing the formula $W=V\times A$, if two factors are known then the third can be calculated as follows:

$$W=V\times A \qquad A=\frac{W}{V} \qquad V=\frac{W}{A}$$

The ohm is the unit of resistance (R) a circuit has which opposes the current. The resistance ohms can be worked out providing the voltage, current or wattage are known, as follows:

$$A=\frac{V}{R} \quad R=\frac{V}{A} \quad V=R\times A$$

The normal formula symbol for amp is I and therefore it will be quite usual to see the above-mentioned formulae written as:

$$I=\frac{V}{R} \quad R=\frac{V}{I} \quad V=R\times I$$

and hence if the wattage needs to be worked out and the current is not known but both the voltage and the resistance are known, then for

$$W=V\times I \quad \text{substitute} \quad \frac{V}{R} \text{ for } I$$

then

$$W=V\times\frac{V}{R}=\frac{V^2}{R}$$

An ohm meter is a piece of equipment which will measure both resistance and continuity of a circuit; most ohm meters also have the facility for checking voltages and current flow (amperes). These meters are relatively expensive and are not an essential piece of equipment for a householder. If, however, you need to check a piece of faulty electrical equipment such as a heating element from a kettle with an ohm meter, you will find that most electrical retailers will check the equipment free of charge.

Entry of Electricity into the House

Electricity normally enters the house along a thick, steel-reinforced, pitch-covered cable from underground. In a few cases the supply is brought to the house by overhead cables, these houses tending to be older types in more rural areas.

This pitch-covered cable then enters a fuse box which is the property of the Electricity Board, and in view of this must not be opened by removal of the lead seal. From this main 60 A fuse two double insulated cables lead to the electricity meter, which again is the property of the Electricity Board and is sealed. Two more cables are taken from the meter to the main switched fuse box or consumer unit.

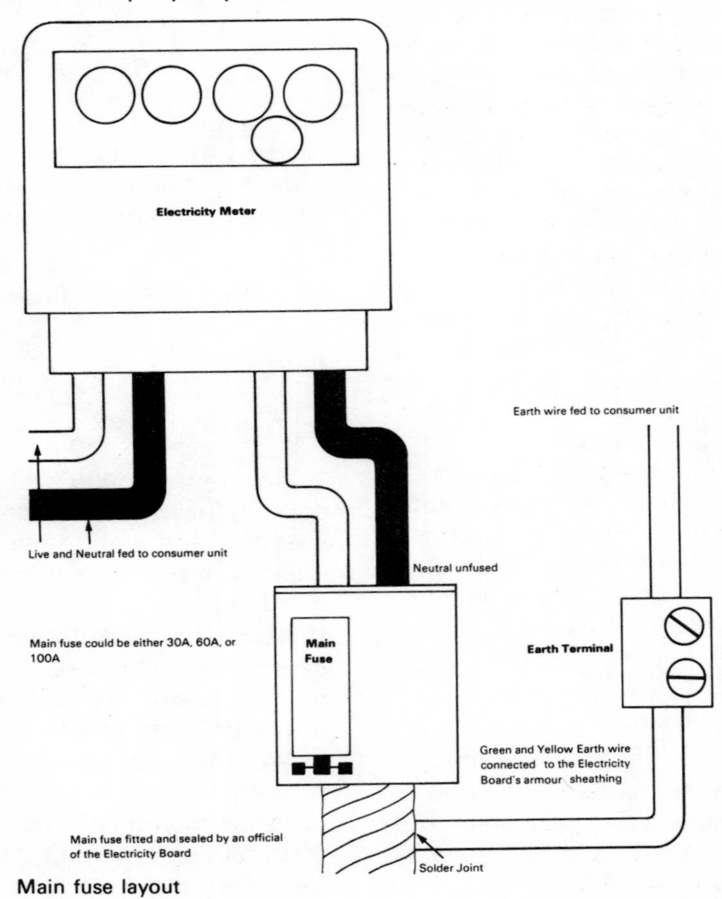

Main fuse layout

In modern houses the main fuse and meter are often contained in a covered recess on an outside wall of the house so that the Electricity Board officials can have access to the contents without needing to enter the house.

Under this arrangement the main fuse box or consumer unit is still contained inside the house. This main consumer unit can usually be found:

- In the cellar—in older houses
- In the porch—usually in a cupboard
- Under the staircase
- In the garage—more modern houses

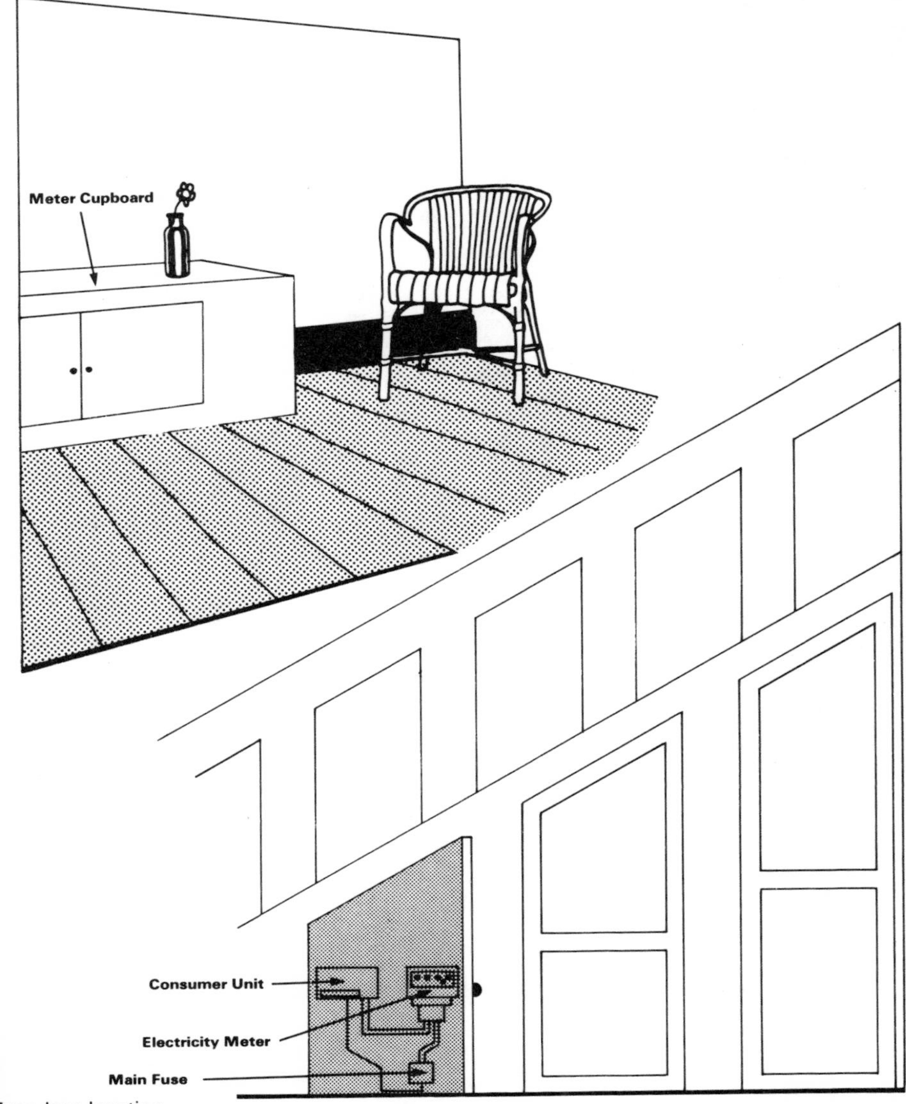

Fuse box location

Precautions

Electricity cannot be seen, heard or detected by smell, but nevertheless can be extremely dangerous, even resulting in fatality. In view of this it is vital that before starting any work involving electricity certain safety precautions are followed.

In all circumstances the **main switch must be turned off.**

It must be noted that in some cases where additions to the original circuit have been made, there may be more than one main switch. An example of this could be in an older property, where an immersion heater has been installed. Owing to the lack of a spur fuse a new fuse box could have been fitted resulting in two main switches. Once familiar with the electrical layout of the house it would not be essential to switch off all main fuses but **if in doubt do switch off all main switches**.

Fuses

A fuse is incorporated into an electrical installation as a safety device to protect the domestic wiring. It consists of a thin piece of wire held in an insulating material. During normal use this allows electrical current to flow uninterrupted. In the event of an overload, a greater flow of current than the circuit was designed for causes the wire in the fuse to heat up and eventually melt. This causes a break in the circuit.

Types of Fuse

In the main consumer unit there are three main types of fuse:

(a) Rewirable fuses
(b) Circuit breaker
} these are found in the main consumer unit

(c) Cartridge fuses—these are found in plugs (and see page 12).

The rewirable types all operate on the same principle and involve the use of fuse wire. This wire is made in a wide range of thicknesses and it is important that the wire appropriate to the fuse is used. Generally, the thicker the wire the greater the current-carrying capacity.

Rewirable Fuses

These are manufactured from an insulating material, usually bakelite or porcelain. The two contact blades (metal pins) are connected by a length of fuse wire, usually held in place by a piece of asbestos.

In the event of an overload the fuse will become blown. The melted fuse will not be in one continuous length and there may be some discoloration of the porcelain or asbestos.

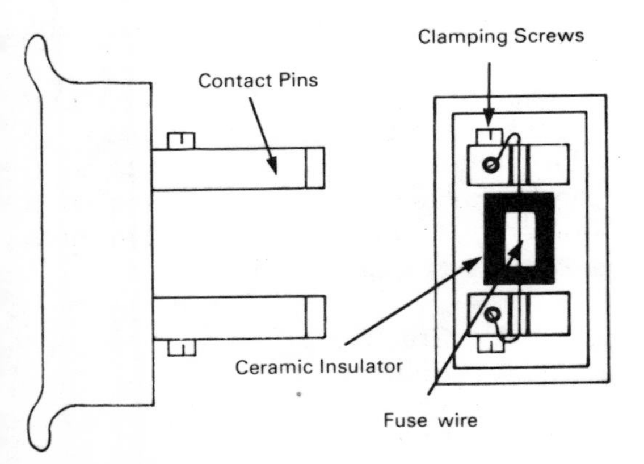

Rewirable fuse

Fuse wire would be broken
There would be signs of burning, and small metal globules of fuse wire contained in the ceramic tube

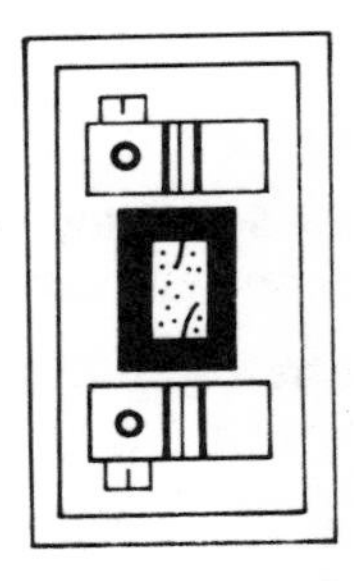

Signs of a blown rewirable fuse

Rewiring a Fuse

1. Switch off main switch.
2. Remove blown fuse from consumer unit.
3. Select correct fuse wire; e.g. for a 30 A fuse use only 30 A fuse wire and likewise for a 5 A fuse use only 5 A fuse wire. The fuse rating is generally marked on the fuse holder.
4. Loosen the screw attached to the contact blades.
5. Remove all traces of previously fitted fuse wire.
6. Loop one end of the fuse wire round one screw and retighten securely.
7. Lay wire across fuse holder in channel provided.
8. Loop the other end round other screw and retighten; *N.B.* do not overtighten screw which may break the fuse wire, or undertighten the screw leaving the fuse wire slack.
9. Cut residual fuse wire close to screw.
10. Replace fuse.
11. Switch on main switch.

If the fuse should blow again there may be a fault in the circuit or in an appliance connected to the circuit and this should then be investigated.

Circuit Breakers

These can be used instead of rewirable fuses and have certain advantages:

1. They break the circuit more quickly.
2. They do not require to be rewired.
3. They last indefinitely.

The biggest advantage in using a circuit breaker is that it does not require rewiring. Should the circuit become broken, the colour-coded button disengages in a push button manner, thus the operation is easily visually identified. Because there is no need to rewire a circuit breaker it is easily reset by simply depressing the button. Should there be a need to isolate the circuit the red button below can be pressed which then disengages the circuit. To complete the circuit the colour-coded button would need to be depressed.

Fuse Ratings

Consumer units can contain up to eight fuse spaces. These can then be fitted with either rewirable fuses or miniature circuit breakers of any value. A typical arrangement could be:

1. Two 30 A upstairs and downstairs power circuits (sockets).
2. Two 5 A lighting circuits upstairs and downstairs.
3. One 30 A or 45 A electric cooker—the 45 A is only required for large cookers.
4. One 15 A or 20 A immersion heater—rating dependent upon type of heater.
5. Two spare spaces for any additions, for example garage, greenhouse, etc.

The rating of fuses and circuit breakers can be easily determined.

In all cases the rating is indicated in the bakelite or porcelain moulded holder. In addition to this, more modern types are colour coded.

45 A	Green
30 A	Red
20 A	Yellow
15 A	Blue
5 A	White

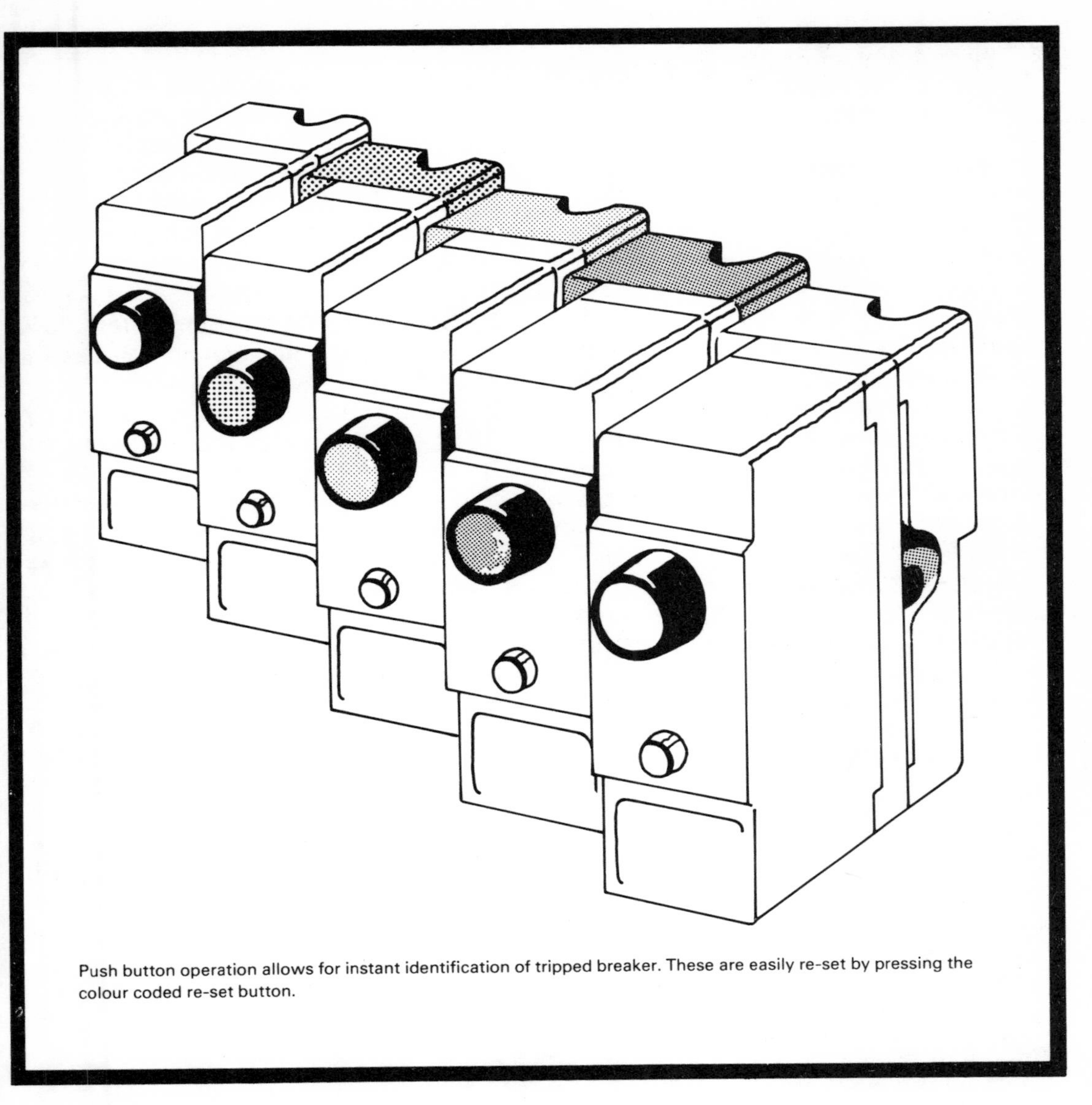

Push button operation allows for instant identification of tripped breaker. These are easily re-set by pressing the colour coded re-set button.

Circuit breakers

Cartridge Fuses

This kind of fuse differs from the rewirable type in that it basically consists of a short ceramic cylinder capped at either end in steel.

The fuse wire is positioned in the centre of the cylinder and is connected to the steel cap at either end. This fuse is not rewirable and if it blows it must be discarded.

As the ceramic cylinder is opaque it is impossible to see if the fuse needs replacing. Therefore, a suitable method must be used to determine if a replacement fuse is needed.

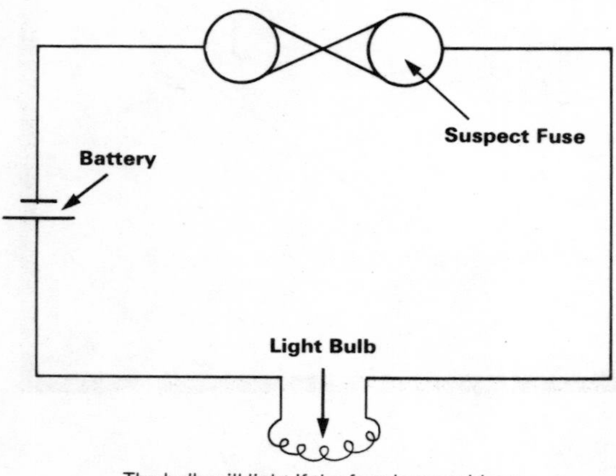

The bulb will light if the fuse has not blown

Fuse-checking circuit

An electrician would use an ohm meter to check fuses, but an alternative method would be to conduct the following simple test:

1. Connect a $1\frac{1}{2}$ V(U.2) battery to a small light bulb such as one used in a torch.
2. Insert the fuse to complete the circuit as shown in the diagram.

If the bulb lights you can be assured the fuse is working. If the bulb does not light the fuse has blown and needs replacing, providing the circuit without the fuse enables the bulb to light.

Cartridge fuses are obtainable in a number of current values to suit varying applications. Typical values are:

1. 13 A—washing machines, electric fires above 1 kw, electric kettles.
2. 5 A—refrigerators, vacuum cleaners, irons, one-bar electric fires.
3. 3 A—electric blankets, hair driers, power drills, sewing machines, food mixers.
4. 1 A—standard lamps, table lamps.

The function of the fuse is to protect the wiring circuit from an overload in the event of an appliance developing a fault and causing an excess current to be drawn.

It is important to check that all manufacturers' recommendations regarding suitable fuse ratings for electrical appliances are complied with.

Calculation of Fuse Rating

To calculate the correct fuse value, a simple calculation is required. The standard voltage obtained from the Electricity Board is 240 V. Each electrical appliance will be rated at a certain wattage.

Using this, one divides the power consumption of the appliance (wattage) by the supply voltage to calculate the current drawn in amps.

for example:

A single-bar (1 kw) electric fire

$$\frac{1000\ \text{W}}{240\ \text{V}} = 4.16\ \text{A}$$

Having calculated the current drawn in amps for an appliance, the fuse with a fuse rating nearest above this value is used.

for example:

A single-bar (1 kw) electric fire

current drawn=4.16 A
fuse rating=5 A

The table below is meant to act as a guide only. Any new electrical appliance should have its wattage noted before any calculation of fuse value is made.

Cartridge fuses are colour coded for easy identification:

1 A	Green
3 A	Red
5 A	Black
13 A	Brown.

Typical current consumption of domestic appliances

Appliance	*Wattage*	*Voltage*	*Current Used*	*Fuse Value*
3-bar electric fire	3000 W	240 V	12.5 A	13 A
single-bar electric fire	1000 W	240 V	4.16 A	5 A
vacuum cleaner	650 W	240 W	2.64 A	3 A
television	260 W	240 V	1.8 A	3 A
food mixer	120 W	240 V	0.5 A	1 A
stereo radio	60 W	240 V	0.25 A	1 A

Cables and Flexes

Rewiring of circuits and electrical appliances requires 'wire' but it is important that the correct type of wire is used. For rewiring circuits 'cable' is used whereas for electrical appliances 'flex' is needed.

Both cable and flex consist of copper wire to conduct the electricity and an outer covering of an insulating material, often plastic or rubber.

Types of Cable

Cable is used for the installation of power and lighting circuits.

This consists of two strands of copper each insulated with plastic and, as these are live and neutral conductors, they are colour coded with red and black plastic sheaths. The earth conductor is not insulated and is positioned between the other two conductors, the whole lot then being surrounded by a grey plastic insulation.

The insulation can alternatively be made of rubber although this does tend to crack and perish more readily than plastic.

It is essential that the correct size of cable is used for a particular application. In general, the thicker the conductor the greater the current carrying capacity of the cable.

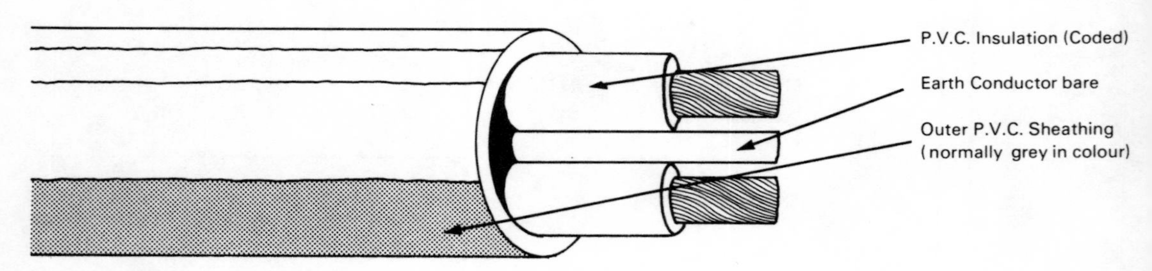

PVC double insulated cable

The measurements in *Figure A* apply to the area of the cut end of the conductor; thus the larger the area of the cut end (cross-sectional area) the greater the diameter of the conductor and so the thicker the wire overall.

All measurements on cables referred to in *Figure A* are metric measurements. Cables other than those recently fitted will not have metric measurements and *Figure B* shows the construction of these.

To identify the type of cable used the copper conductors can be examined to determine whether they are single-stranded or multi-stranded conductors, as shown in the diagrams above.

It is possible to connect up to two types of cable as they have the same current-carrying capacity, e.g. 3/029 cable can be safely connected to 1.5 mm^2 cable.

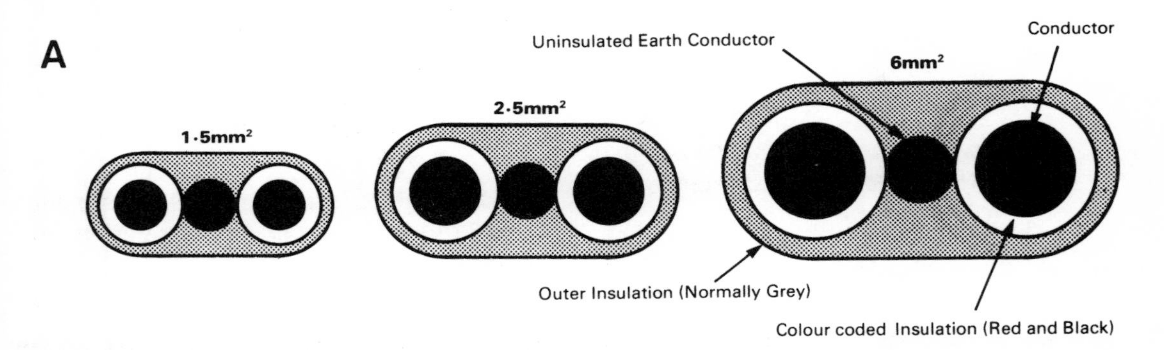

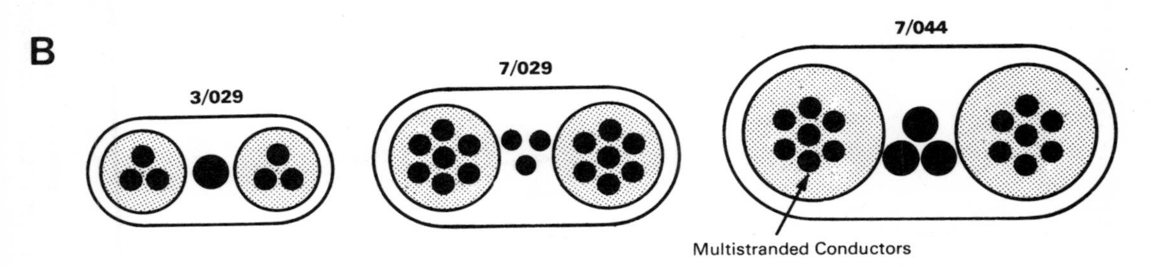

Cross-sectional view of cable sizes

Types of Flex

Flex is used for connecting electrical appliances to the socket outlet.

Two-Core Flex: This can be obtained in various forms:

(a) *Twin-core flat*—two strands of non-insulated wire moulded into a plastic insulator.

Colour code—none. The plastic insulation is manufactured in a range of different colours. There is therefore no significance in the colour of the insulation as there is in the case of the cables already detailed.

(b) *Twin-core insulated*—two separate strands of insulated wire twisted together.

Colour code—none. This again is manufactured in a range of different colours and in this respect is similar to the previous flex.

(c) *Twin-core circular*—two insulated strands surrounded by an outer insulation.

Colour code:
Live conductor brown
Neutral conductor blue.

On older appliances the flex colour code is:

Live conductor red
Neutral conductor black

These are surrounded in an outer insulation which can be in a variety of colours.

These two-core flexes are used on

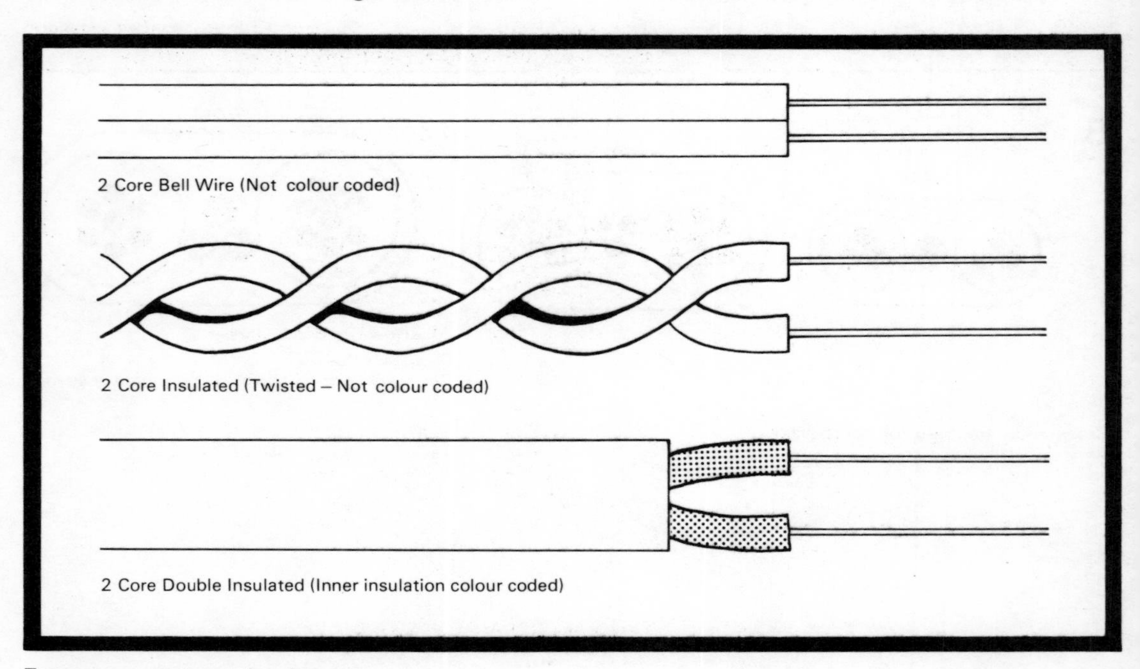

Two-core current-carrying flexes

low power consumption appliances where no earth wire is required. **On no account** must twin-core flex be used on metallic or part metallic appliances.

Twin-core flex must only be connected to the live and neutral terminals of the plug and **under no circumstances** must the earth terminal be used.

Three-Core Flex: This can be obtained in three forms:

(a) *P.V.C. double insulated*—three multistrand strands of wire each insulated in colour coded plastic or rubber surrounded by an outer plastic or rubber insulation.

(b) *Butyl rubber double insulated*—this is similar in construction to that just described, but differs in that all the insulation is of heat resisting butyl rubber.

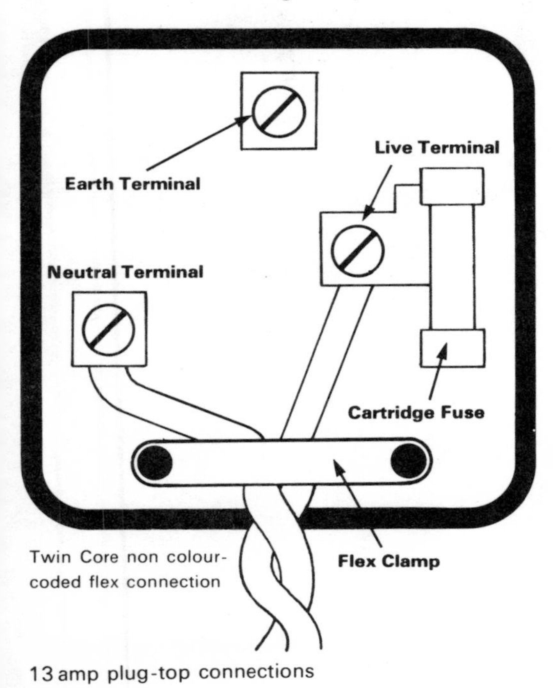

13 amp plug-top connections

(c) *P.V.C. double insulated* (*non-kink*)—this is similar to the previously described three-core flex but differs in that it contains a layer of separate strands of fibrous material contained inside the outer insulation. This gives the flex a greater rigidity and prevents knotting and tangling.

All three-core flexes are manufactured with the conductors insulated with colour coded plastic and rubber.

Colour code—

Live conductor	brown
Neutral conductor	blue
Earth conductor	green and yellow striped.

On older appliances the flex colour code is:

Live conductor	red
Neutral conductor	black
Earth conductor	green.

Three-core flex must only be connected with the earth conductor connected to the earth terminal of the plug.

Ratings and Uses

Two-core flex is used only for low current consuming appliances, where no earth connection is required.

The thickness of the flex is determined by the current consumption of the appliance. Therefore, an appliance drawing a large current requires thick cable in comparison to one drawing a small current which requires a thin cable. All types of twin-core flex are available up to 6 A current carrying capacity. Two-core cable with a current rating from 3 to 6 A can be used for:

electric clocks, hair driers, table lamps, standard lamps, some portable hand lamps, christmas tree lights and all double insulated domestic appliances where no earth is required.

Three-core flex can be used for all appliances where an earth wire is required. It is also capable of carrying large currents ranging from 3 A to 20 A or even more in certain cases. It is important when connecting an appliance to a socket outlet that the correct current carrying capacity flex is used.

Typical applications are:

Electric kettle	Washing machine
Refrigerator	Electric drill
Vacuum Cleaner	Dish washer

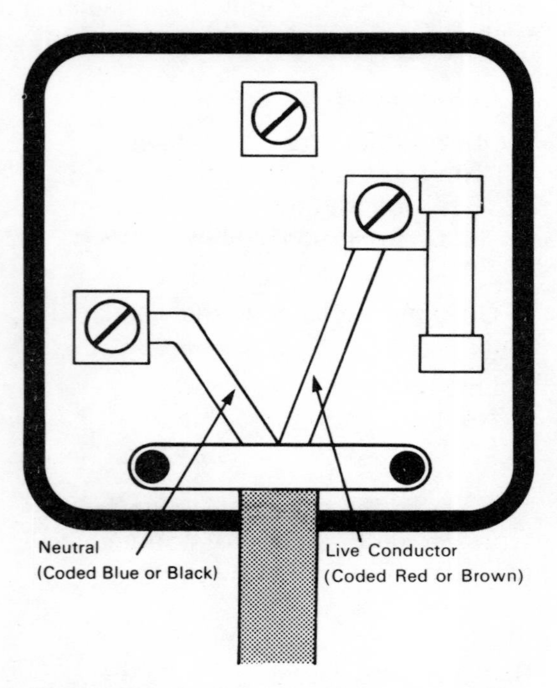

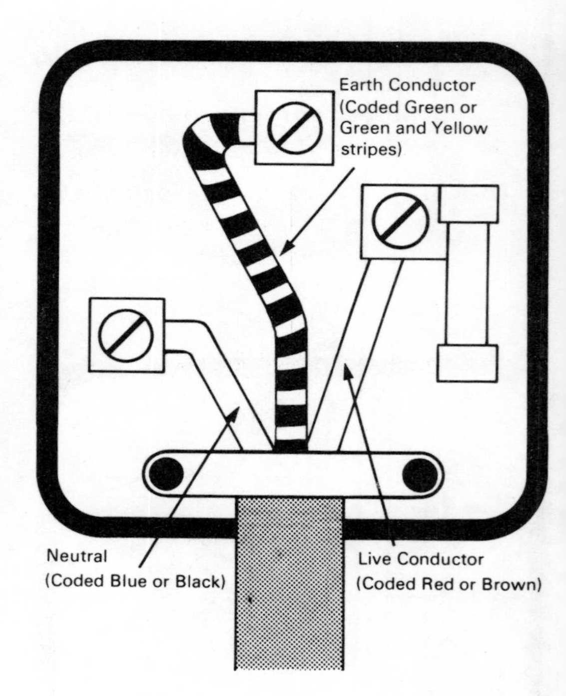

Plug connections for colour-coded two-core and three-core flex

Plugs and Sockets

Plugs

There are a variety of different styles of plug but the basic requirements are all the same. Plugs are manufactured to a British standard specification (BS 1363). The most generally used plug is the white bakelite type, the later styles of which have semi-insulated pins that prevent small children accidentally being electrocuted should they ever be allowed to come into contact with them.

At the base of the plug is the cord grip which in all cases must clamp the outer sheathing of the flex; failure to do this could result in the cable becoming detached from the three terminals and a serious short could occur. The three terminals used to clamp the conductors are normally one of two types.

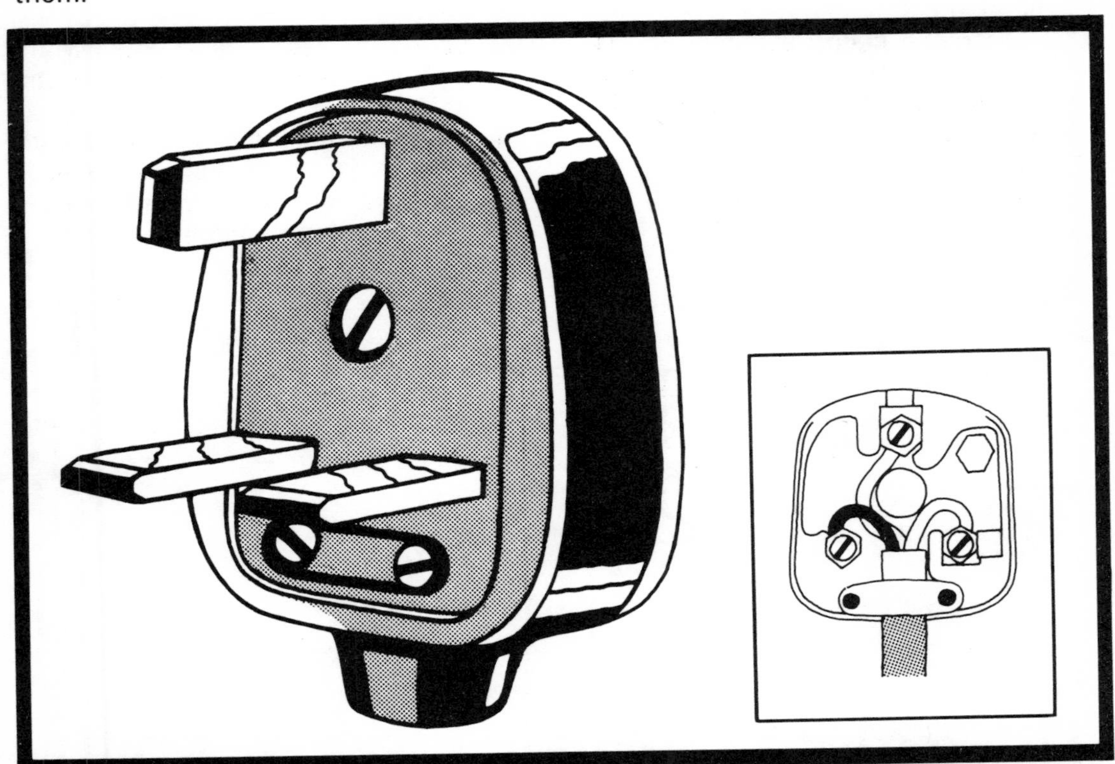

13 amp plug

1. The terminal has a hole in it which the wire passes through and is then clamped securely in place by the locking screw.
2. The wire is clamped to the terminal by a knurled, slotted nut.

There is no particular advantage or disadvantage with either type, but in the case of the latter the cable must be twisted round the threaded terminal in a clockwise direction—failure to do this could result in the cable becoming detached as the screw is tightened. The best method of wiring a plug is:

1. Strip back 50 mm of the outer sheathing of the flex.
2. Pass the flex through the cable clamp until the flex is directly under the clamp.
3. Tighten the clamp securely. In the case of a rubber crush-proof plug the plug top must be slid over the flex before connection to the plug takes place.
4. Connect the live conductor to the fused side of the plug taking care to trim back the flex in order to leave no slack.
5. Connect the neutral conductor in the same manner.
6. Connect the earth conductor.
7. Check fuse rating of the appliance fitted to the plug and fit accordingly.
8. Replace plug top.

Sockets

There are many types of sockets each having a particular application.

Three pin square sockets are used for both domestic ring main circuits and industrial applications. The socket face can be manufactured from either:

- White bakelite (general domestic use)
- Satin stainless steel with white plastic inserts (general domestic use)
- Aluminium painted mild steel with plastic inserts (industrial applications, garages, etc., where appearance is not paramount).

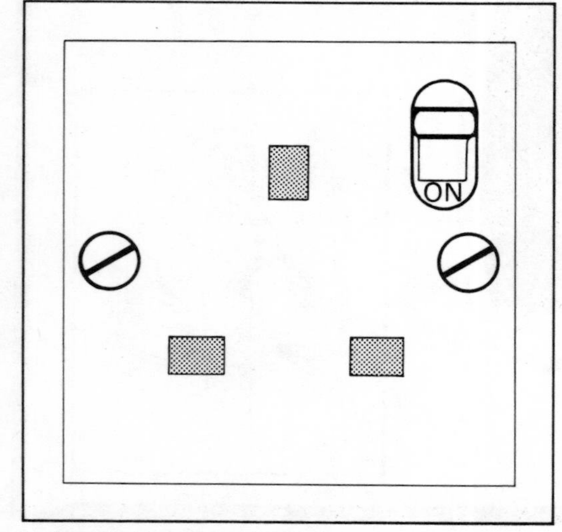

3 pin switched socket

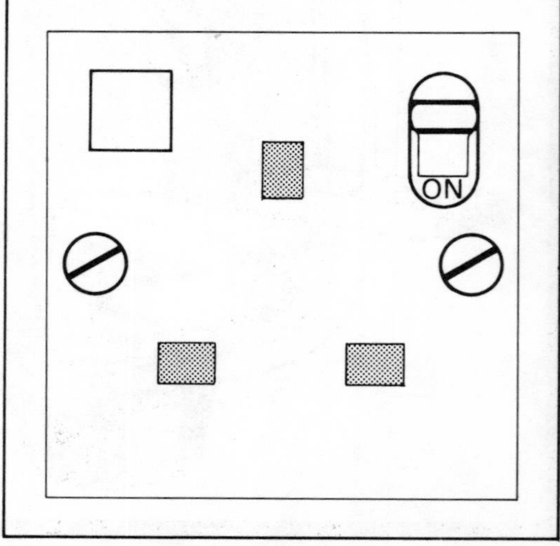

3 pin switched illuminated socket

The socket can be mounted flush to the wall or raised 50 mm from the surface. The two methods of fitting are known as 'flush fixed' and 'surface fixed'. For most industrial applications, e.g. factories and laboratories, sockets are surface fixed, that is to say a steel box is screwed on to the wall and the socket face screwed on to the front of this. The cable is fed to the socket by passing it through conduit tubing.

For domestic purposes plastic boxes are obtainable for surface fixing sockets; these are neater in appearance than the industrial steel type. The cable is 'chased' into the socket, which means that the wall plaster is removed and the cable laid in a channel. The cable is then covered by a plastic cover (capping) to prevent it being accidentally pierced. The channel is then plastered over. An alternative method is to flush-fix the socket. For this a hole needs to be made some 75 mm × 75 mm and approximately 50 mm deep to allow the box to be fitted flush with the wall. The box can either be held in place with screws and plugs or cemented in. The socket face can then be screwed to the box which will fit flush with the wall.

Both single and double sockets are obtainable; these can be either switched or unswitched and in some instances are fitted with a neon light. It is preferable, and far safer, to fit a double socket rather than to fit a single and risk the possibility of overloading it by inserting a two- or three-way adaptor, sometimes called a jack, to increase the number of appliances used from the socket.

Sockets with an inbuilt neon light are ideal where an electric kettle is used so that an immediate visual check can

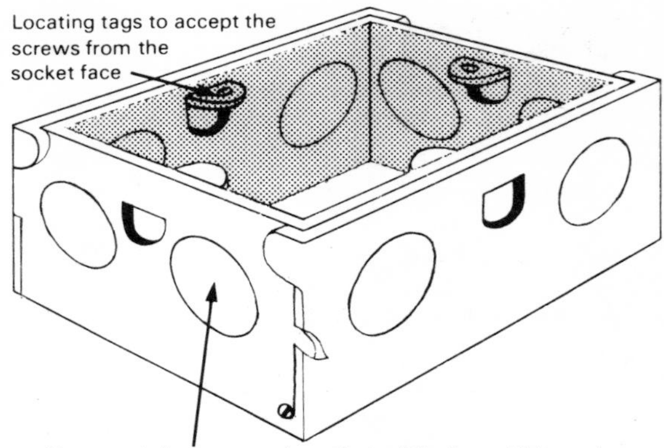
Locating tags to accept the screws from the socket face
The round discs are weak sections of the box which can be nicked out to allow entry for the cable
Wall-depth socket box

be made to see whether it is switched on or off.

Cord outlet fused sockets are available for certain applications. These consist of a plain socket face with the facility for a special fuse carrier to be fitted. Instead of the outlet consisting of three pins into which the plug fits, a piece of flex is fed through a hole in the front and connected internally, this flex in turn being connected to an appliance. Two typical applications for this type of socket face are:

- Central heating boilers
- Night storage heaters

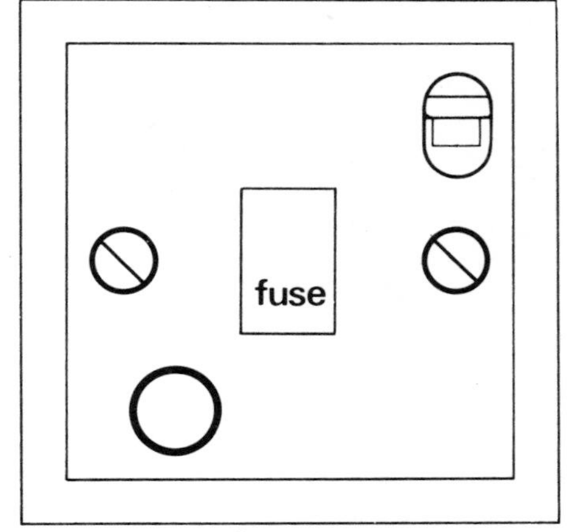
fuse
Cord-outlet fused socket

Switches

Fitting and Removal

Modern switches are held to the wall by two 2 BA screws and are usually white in colour. Single switches are generally the most common and can be obtained in as many as up to four in a block. Quite often single on/off switches are made with a two-way switch in one unit. A typical application for these is in a hallway where a single-way downstairs light is fitted and an upstairs two-way light is used. This allows for the upstairs light to be switched on from downstairs.

To surface fix a switch to the wall a plastic box must be purchased. This box is similar to the plastic box used for surface mounting a socket but is only some 15 mm deep. The reason for this is that the cable used in lighting circuits (1.5 mm²) is thinner and that therefore it does not take up so much room.

To flush fix a switch a steel box needs to be set into the wall similar to that of the flush-fixed socket. Again this is only some 15 mm deep and usually only requires the plaster to be chipped from the wall to allow it to fit flush with the surface.

To connect a single on/off switch it will be seen that there is one piece of cable (usually 1.5 mm²) entering the box in which there are live, neutral and earth conductors. The earth conductor is connected to a screw usually mounted in the back of the box, whether the box be made of plastic or steel.

To connect the live and neutral conductor, there are two terminals on the back of the actual switch, some have 'live' marked against one terminal and 'switch' against the other. In this case the red conductor should be connected to the 'live' and the black to the 'switch' terminals. It is not essential, however, to do this as some switches are not marked. However, if the wires are changed over, the switch's 'off' position will then be in the normal 'on' position. 'Off' is usually the upward position, 'on' the downward one.

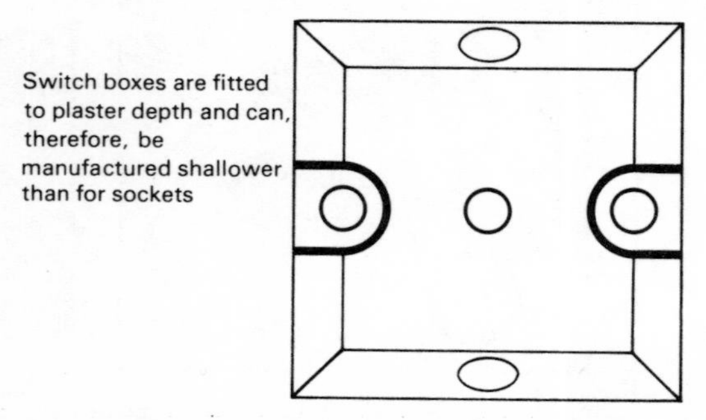
Switch boxes are fitted to plaster depth and can, therefore, be manufactured shallower than for sockets

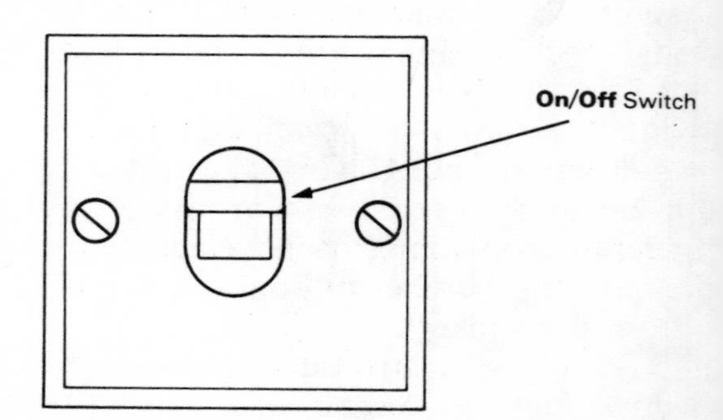

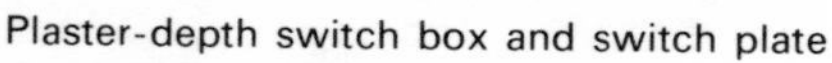
Plaster-depth switch box and switch plate

Dimmer Switches

Dimmer switches can normally replace any single on/off switch. There are only two connections, similar to those of a normal switch. Dimmer switches are an electronic device used to limit the current and hence produce the effect of dimming the light. It is important when purchasing a dimmer switch that the power handling capabilities (wattage) are noted as these vary from manufacturer to manufacturer.

Some dimmers have an on/off switch fitted to them, the advantage of this is that when the correct brightness has been selected the dimmer can be switched off at this setting. On re-entering the room a level of instant light is obtainable by merely flicking on the switch.

In the other type of dimmer the switch is incorporated in the variable light control knob. These controls therefore require to be switched on and the brilliance increased gradually. Illumination is not as instantaneous as in the other type but the switch does look slightly neater.

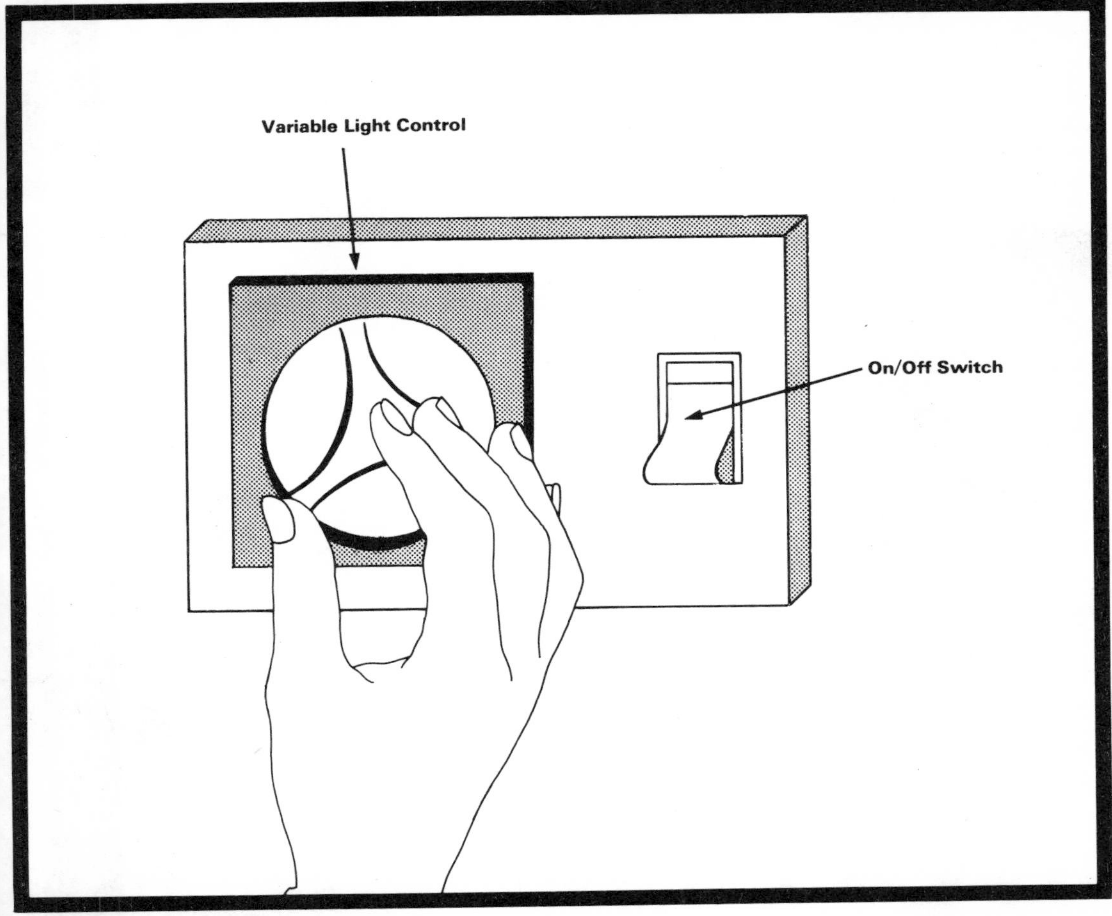

Switched dimmer control

Two-Way Switches

The fitting of one of these is similar to an ordinary one-way switch apart from the fact that there are three terminals on the back of the switch, two of which are marked 'common'.

When fitting a two-way switch a piece of 1.5 mm² twin-core cable with earth must be used to connect the two switches together. A live feed must be fed to one switch and a neutral taken from the other switch to the ceiling rose.

In a great many cases a two-way switch is mounted with a single-way switch as used in a hall. The live conductor can be looped from the live side of the single-way switch to the two-way switch. From the upstairs switch a neutral conductor must be taken to the light rose and the lighting flex must be connected to the two neutral wires.

It must be noted that a two-way switch will not always be 'off' in the downward position or in fact 'on' in the upward position.

An alternative to this is to use a piece of 1.5 mm² three-core cable with earth to connect the switches together. The coding of the cable is red, yellow and blue. The red is connected to common, and the yellow and blue to the other terminals. The red lead from one switch is connected to live, the red lead from the other is connected to the lamp, the other side of the lamp being connected to neutral. All connections are made through a six-way junction box.

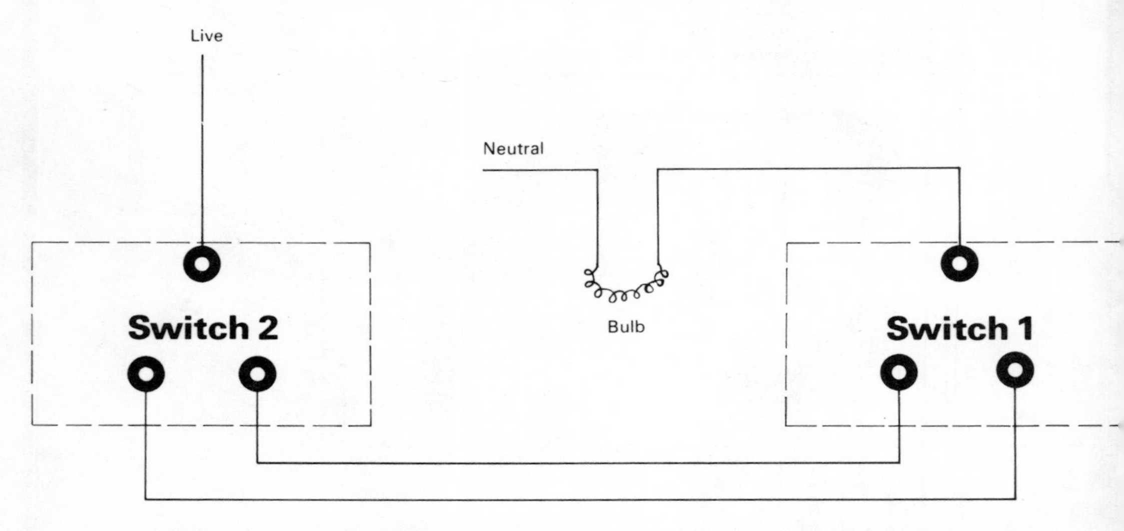

Two-way switch wiring circuit

Pull-Cord Switches

These switches are fitted to the ceiling and are operated by a pull cord. It is essential that **in a bathroom this type of switch is always fitted.** A normal wall-mounted switch contravenes the regulations governing electrical installations. Pull cords are often used in bedrooms in a two-way switch configuration which allows one to switch the light off or on whilst in bed.

Some pull-cord switches light up when switched on; these are mainly used in bathrooms where a light/heater is fitted and this gives an indication that the heater is turned on.

The first job in fitting a pull-cord switch is to secure the back plate to the ceiling. This means that the ceiling joist must be found and the plate screwed to it; an alternative to this is to secure a piece of wood between the joists and fix the plate to this.

Once the back plate has been fixed the switch is screwed to it with two 2 BA screws. If one decides to fix to a joist instead of to a piece of wood fitted between the joists, it should be remembered to fit the plate to one side of the joist so as to allow the cable to be fed through to the switch.

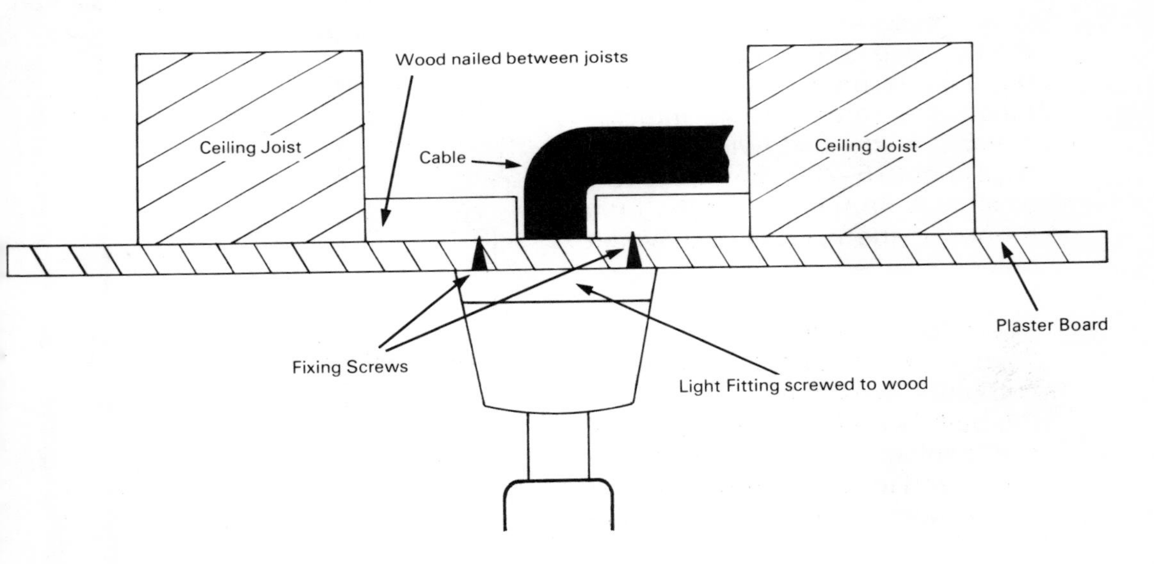

Ceiling rose fixing

Bulbs

Types

There are two different types of fittings:

(i) Bayonet Type
(ii) Screw Type

(i) The bayonet type is the more common and is generally found in the everyday household. This consists of a plastic holder with two spring-loaded brass contacts which make the connection with the two lead contacts at the end of the bulb. In order to fit a bayonet bulb it has to be pushed into the bayonet holder and twisted clockwise about 15–20 degrees, the two small lugs on the side of the bulb fitting into the insets of the bayonet holder.

(ii) The screwed type is similar to the bayonet type externally but is fitted with an internal thread into which the bulb is screwed. There is only one contact in the centre of the holder and the brass cap on the end of the bulb serves as the second contact.

Until recently domestic bulbs have been rather unimaginative and were only different from one another in their wattage ratings, i.e. light output. These range from 30 W to 150 W and can still be purchased with either clear or pearl (opaque) glass. During the last few years bulb manufacturers have started to produce more different and unusual colours and types of lights. Bulbs can be designed to give a wide range of colour and different types of beam, i.e. floodlight, diffused beam and spot concentrated beams, and the majority of these can be obtained in either bayonet or screwed types. The more exotic bulbs are usually used with a track lighting set up—this is explained more fully on page 28.

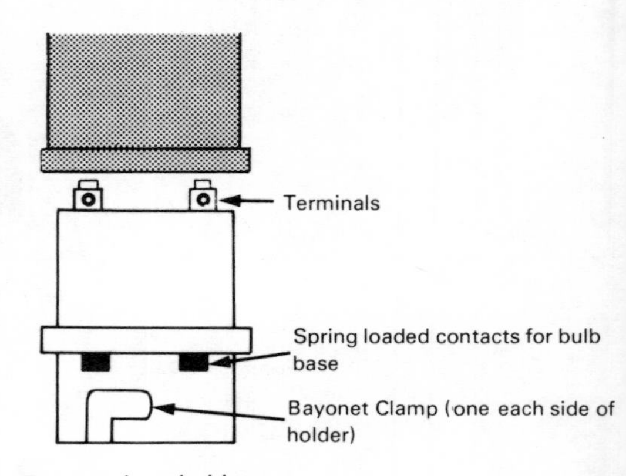

Bayonet lampholder

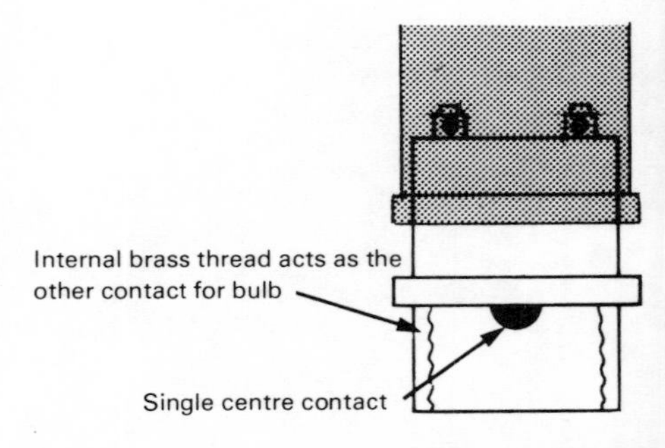

Screwed lampholder

Fluorescent Tubes

These tubes are made of glass and fitted with two steel end caps from which protrude two small pins. Fluorescent tubes are bought in length and not wattage rating as in the case of ordinary bulbs. There are, however, different types of light given off from a variety of tubes and an example of this is 'warmlight'. This type of light is the normal one for domestic use. Another type is 'colour match' which gives a light as near as possible to actual daylight. This latter type can, of course, be fitted into any normal fluorescent light but is more generally used where a very natural, harsh light is required, for example in a spray booth where extreme accuracy is needed to match paint.

In addition to normal straight tubes there are also round and U-shaped ones. The round tubes, when suitably concealed behind a shade, make an ideal light for a lounge where a good light is needed. The U tubes can be used where concealed lighting is fitted behind curtains, pelmets, etc.

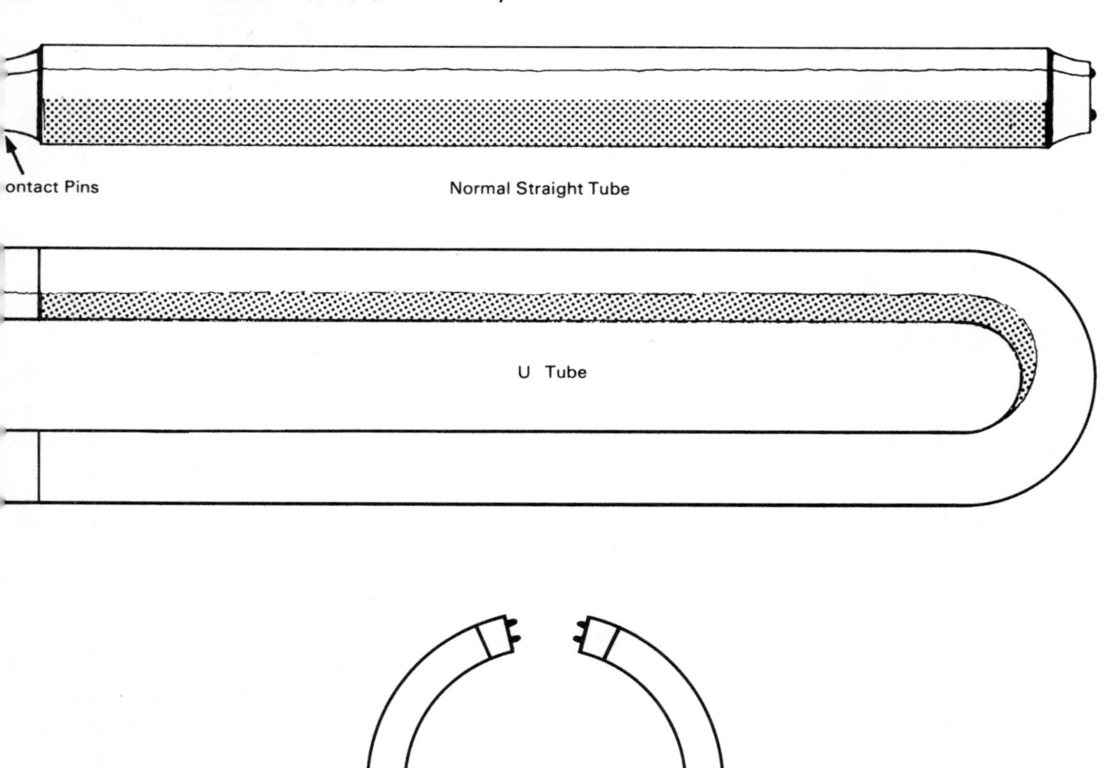

luorescent tubes

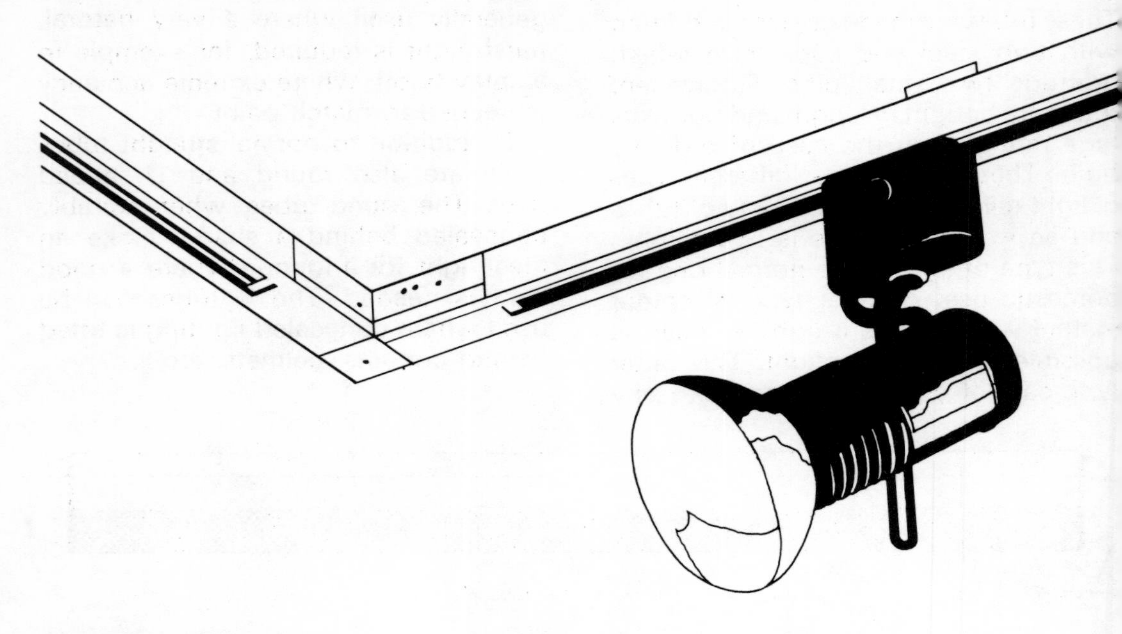

One type of track lighting

Track Lighting

This is a modern concept of lighting and can be installed reasonably easily in a normal home. Some manufacturers give as many as four individual circuits in one track and therefore by arranging the different types of light, i.e. spot, diffused and the varying colours, a good effect can be obtained.

The track can be fitted from an existing ceiling rose if desired, but it must be borne in mind that if this method of connecting is used all the lights on that section of the track would operate simultaneously, but if four separate wires are fed from switches then there can be many permutations of lighting. Electrical appliances with a power rating of less than 500 W can normally also be connected to a track system.

To realise the full versatility of a track system it is probably ideal to consult the literature produced by the various manufacturers before installing a track. It must also be appreciated that each individual manufacturer's specifications differ slightly from the others.

Christmas Tree Lighting

All Christmas tree lighting operates on a two-wire system, that is, no earth wire is connected. Two basic types of system exist:

(a) Series wired
(b) Parallel wired

(a) *Series wired*. This means that if one light bulb is taken out of a socket then the complete light set will be extinguished. If, for example, there were 24 bulbs in a set then the voltage required at each bulb would be 10 V, i.e. 240 ÷ 24—this is the most common type of configuration.

(b) *Parallel wired*. This means that if one bulb is omitted from the circuit the other bulbs will still operate and the working voltage of each bulb would be 240 V.

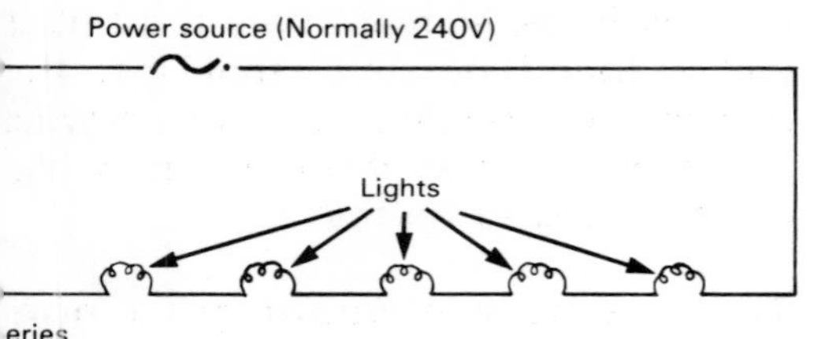

eries

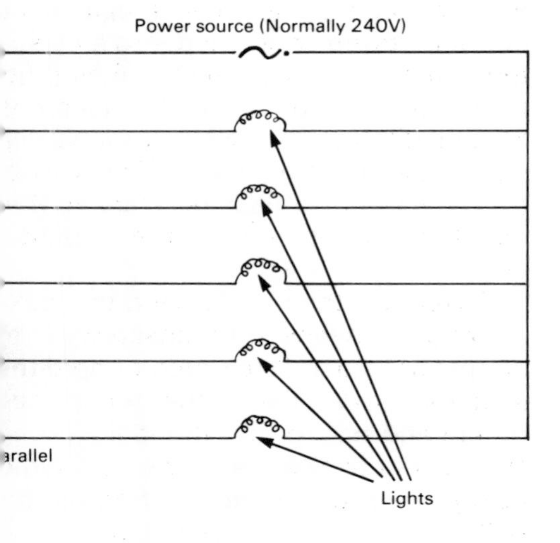

arallel

eries and parallel wired circuits

Generally, when Christmas tree lights are brought out of the attic a couple of weeks or so before Christmas they do not always work. If this should be the case the first thing to do is check the plug connections (i.e. live and neutral) and also the fuse and its rating. If these are found to be satisfactory a common complaint is usually that the bulbs are loose in their sockets and therefore are not making a good contact in the bulb holder. If, after checking the bulbs, they still do not light up it could be that in the case of the parallel wired all the bulbs have blown (although this is most unusual) or that the wire has broken somewhere along its length. If this is suspected then each conductor can be checked for continuity by disconnecting from the mains and then using an ohm meter to check continuity, or the bulb and battery method can be used as described on page 12.

In the case of the series connected lights, checking for continuity between each bulb holder can be carried out in the same manner as for the parallel connected ones. However, to check the actual bulbs may be more difficult and ideally the most satisfactory method is to use an ohm meter or, if the coloured glass is not too opaque, a visual check can be made. Failing this, check the bulbs' working voltage and then connect the batteries together to obtain the voltage, following which the bulbs should be checked. A car battery (12 V) is quite often sufficient for this method and in a great many cases is easily obtainable as the majority of householders own a car.

Electrical Appliances

Kettles

Electric kettles contain a small electrical heating element in the bottom of the kettle. The live and neutral conductors are encapsulated in a steel sheath containing a white powder which acts as the insulation. Most kettles have a rating of 2.5 to 3 kw which is equal to a three-bar electric fire. These elements sometimes break. The usual cause for this is:

1. Switching the kettle on without ensuring that there is sufficient water in it to cover the element.
2. Insulation breaking down, thus causing a short circuit.
3. In some hard-water areas it will be found that the elements are sometimes attacked by the salts in the water. These salts cause the outer steel sheathing to corrode and ultimately allow water into the insulation, causing a short circuit.

When purchasing a kettle it is important that the appliance conforms to a British Standard Specification. Most of the British manufacturers construct their kettles to conform to this specification and in most cases are better than the specification demands. The normal kettle will require to be switched off once the water has boiled and therefore it could be worth bearing in mind that a 13 A plug which has a red neon light fitted to the plug top might serve as a visual check when the kettle is switched on. Most modern kettle elements have a fail-safe device which ejects the plug connecting the kettle to the flex if the element gets too hot. It will be found, however, that once this device has operated it is not always resettable and may result in a new element having to be fitted.

To fit a new element to a non-automatic cut-out kettle

The element consists of the conductors encapsulated in a steel sheath. The conductors are connected into a plastic termination (A) into which the plug fits to conduct the power to the element. Moulded to this is a steel sleeve which has a thread cut on it (B). The washer (C) is made of a heat-resistant rubber which fits tight against the shoulder (D).

To remove the old element, the locking ring (E) needs to be unscrewed; to do this one hand is required to hold the element inside the kettle whilst the other hand has to turn the locking ring (E) **anti-clockwise.** Try to avoid using a wrench to grip the ring as this will scratch the chrome plating.

After the locking ring has been removed the element can be taken out from the inside of the kettle. It will be noted that the rubber washer will be tight against the shoulder (D)—this forms a water-tight seal to prevent

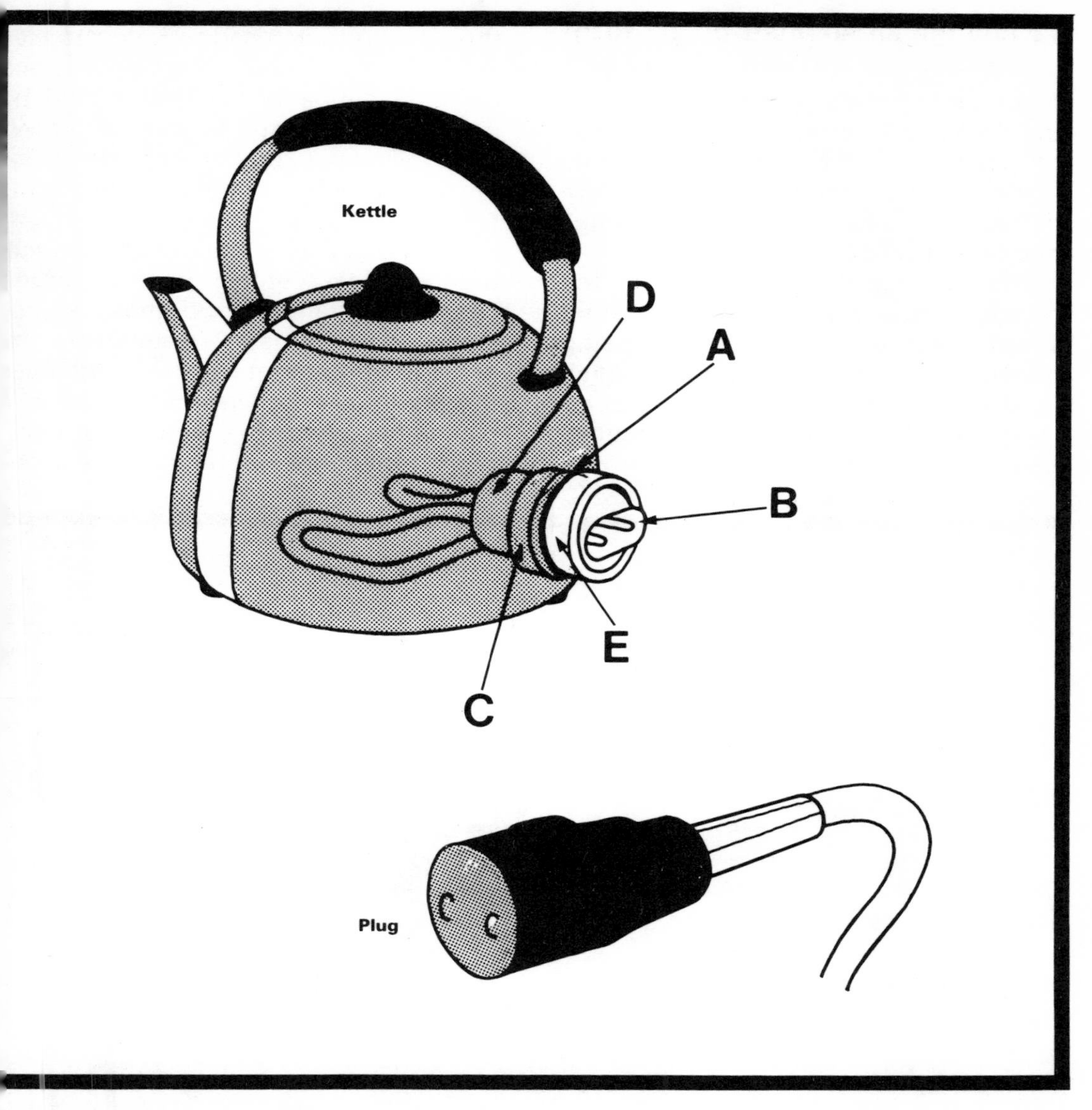

lectric kettle and connecting plug

vater from leaking from the kettle. Different brands of kettles tend to make the collar (B) of varying diameters so it is mportant, when purchasing a new element, to point out to the supplier which brand of kettle you own. Assuming the element is the correct type, place it hrough the lid of the kettle ensuring that the rubber sealing washer is in place. Insert the threaded collar through the hole in the back of the kettle and screw on the locking collar, applying a firm pressure while holding the element steady in the kettle. Fill the kettle with water and check for leaks before connecting the flex and switching on.

To change an element on an automatic cut-out kettle

There are now a number of kettles on the market which automatically switch off when the water boils. The elements in these kettles tend to be slightly more difficult to change than the ones mentioned previously.

An automatic kettle works on the principle of a metal plate expanding when it heats up due to the steam being generated; this pushes a plunger which in turn breaks the electrical contact.

The element differs from a normal electric kettle inasmuch as there are three fixing bolts used to hold the element. Two of the bolts are held in place by brass nuts and the third is held by a round slotted nut (a special screwdriver with the centre of the blade missing is needed in order to tighten this screw).

The points which are used to break the electrical contact become pitted after continual use and these sometimes stick, ultimately resulting in the kettle not cutting out. If this is the case the points can be cleaned by using a piece of emery cloth to remove any pitting; if this does not remedy the prob-

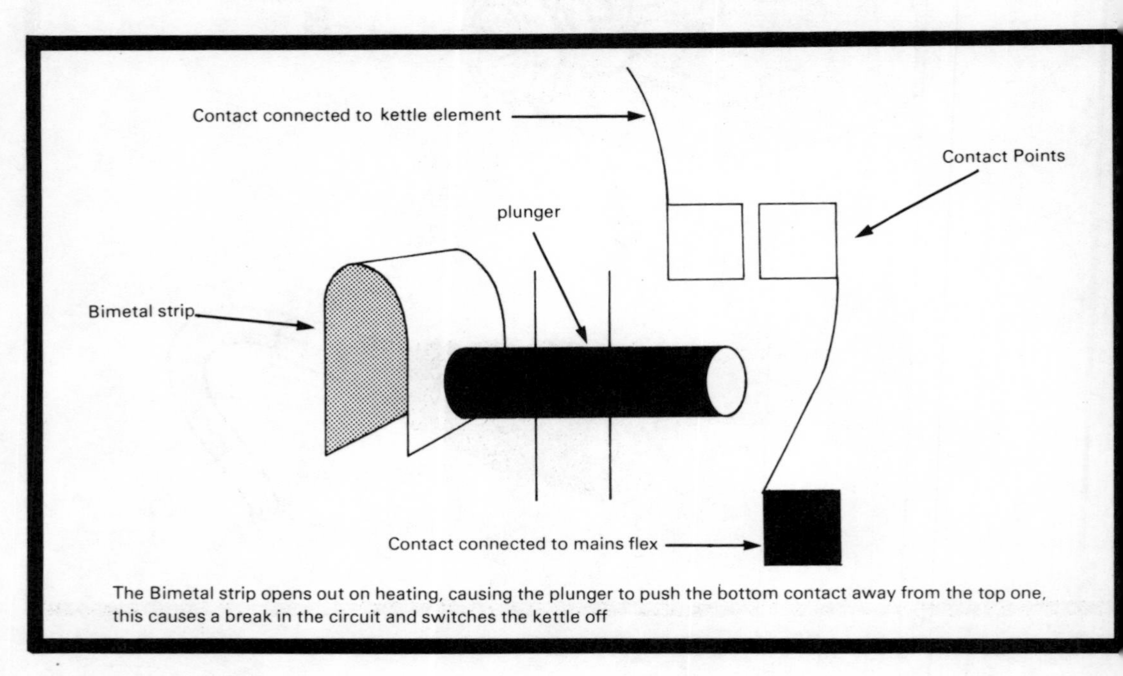

The Bimetal strip opens out on heating, causing the plunger to push the bottom contact away from the top one, this causes a break in the circuit and switches the kettle off

An automatic kettle cut-out

lem a new pair of points can be fitted and in some cases it may be beneficial to fit a complete new cut-out assembly, which is not too expensive and can normally be obtained from an electrical retailer. It will quite easily be seen how to fit this new assembly by comparing the relative holes in the new one with the one already fitted in the kettle; and by simply removing some six screws one is able to fit the new assembly.

One important item to remember when using an automatic kettle is never to leave the house unattended with the kettle switched on. It is a rare occurrence to hear of a kettle malfunctioning but **it can happen** and this will result in the house becoming full of steam. Damage to the kettle element will also occur as in most cases there is no plunger fitted which ejects the plug connection to the kettle, as in the non-automatic cut-out type.

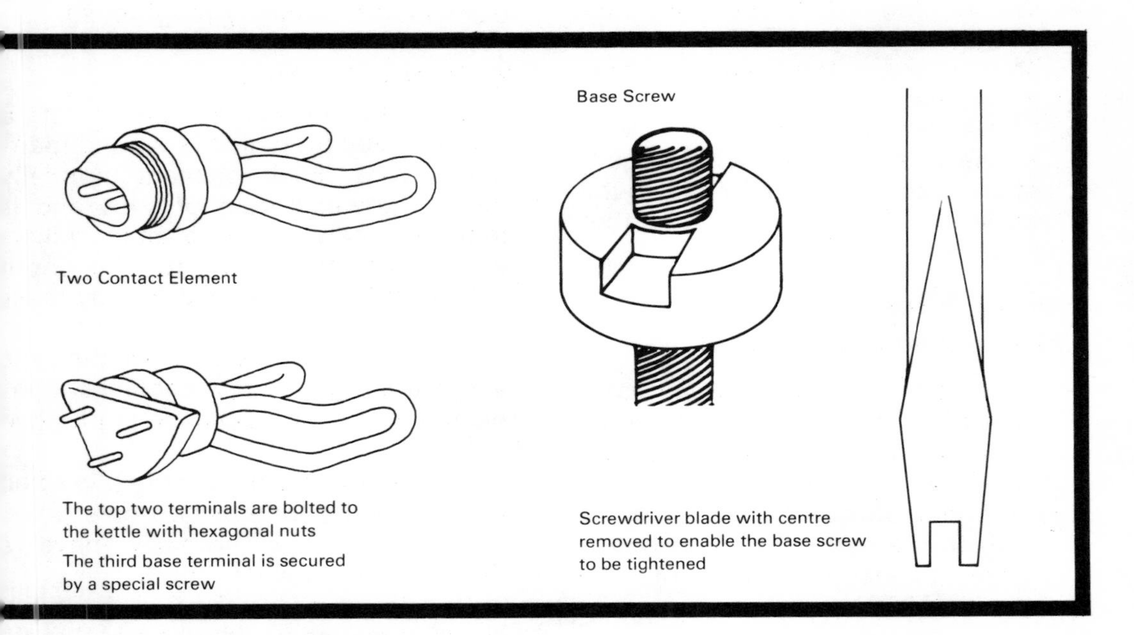
Two Contact Element
The top two terminals are bolted to the kettle with hexagonal nuts
The third base terminal is secured by a special screw
Base Screw
Screwdriver blade with centre removed to enable the base screw to be tightened

Kettle elements, base screw and modified screwdriver

Fault Diagnosis

Kettle failing to heat water.

1. Check fuse—if faulty replace. If the new fuse blows there is possibly a short circuit.
2. Check flex to see if the insulation has become worn and shorted the two conductors together; if this is the case **do not repair with insulation tape—replace it**.
3. Check flex to see if there are any breaks in the wire conductors.
4. Check plug which connects the kettle to the flex. Take particular note to see if there are any burn marks, etc.
5. Check element with an ohm meter. If one is not available then quite often an electrical retailer will do this for you. If found to be faulty then replace.

Electric Fires

There are two basic types of fire:

(i) The convector heater which is an electrical heating element with a fan fitted beneath it to help spread the heat into the room.

(ii) Chrome reflector heater in front of which there is either one, two or three elements, usually of 1 kw each (1000 W).

Convector Heaters

These usually have a spring coil element supported on ceramic insulations. A dial is normally found on the front of such an appliance and this regulates the heat output. The fan works continuously while the appliance is switched on. Some heaters have a red light fitted on to them to give a red glow, which does little to the fire's efficiency but gives it a more pleasing appearance when switched on. When fitting a new element it is important to make sure the correct one is purchased for the appliance. To replace an element normally requires only two or three securing screws to be removed together with the connections to the conductors. When the element has to be changed make certain that the screw terminals holding the live and neutral conductors to the elements are securely retightened.

Reflector Heaters

The heater elements consist of a ceramic tube on which is wound manganin wire. Each element normally gives a power output of 1 kw and is mounted between two steel pillars which are connected to the input mains, and these are held firm by brass nuts.

If the fire has only one bar then the element automatically becomes connected when the appliance is plugged in.

With a two-bar fire a switch is fitted to select either one or two bars.

Three-bar fires normally have a three-position switch fitted.

Special types of wall-mounted electric fires can be fitted in a bathroom but these need to have a special type of element fitted. The value of the element is 750 W—i.e. smaller than a normal fire—and this has to be encapsulated in a silica glass envelope to protect the element from steam and water and to prevent a short circuit occurring.

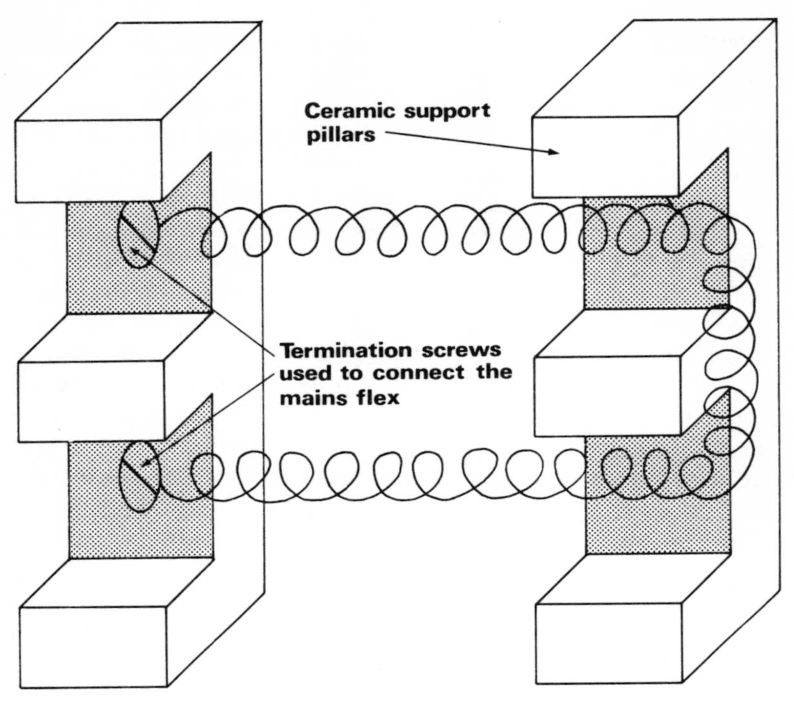

Open-wound convector heater element

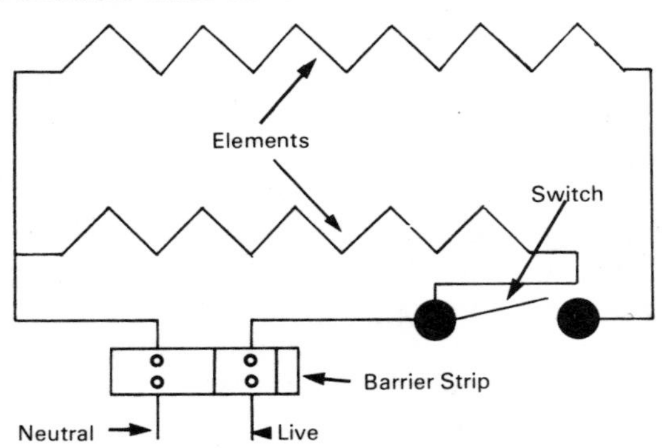

Element connections of a two-bar electric fire

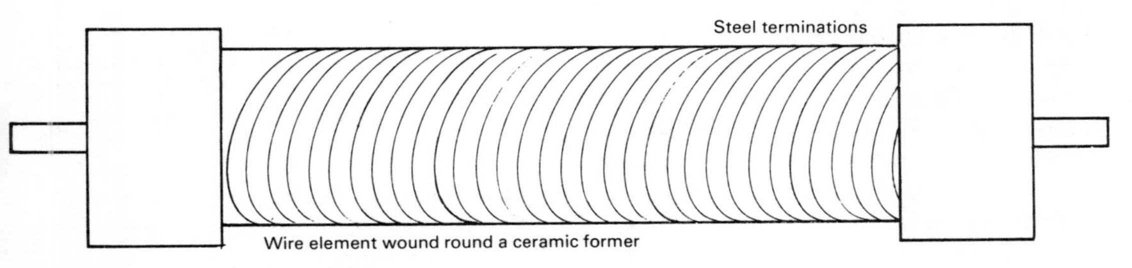

Close-wound reflector heater element

Electric Cookers

How to install an electric cooker

In the majority of kitchens there is an electric point marked 'cooker' which usually incorporates a switched socket, and this can also be used for various other kitchen appliances. The switch on the cooker panel marked 'cooker' will isolate the supply to the cooker. The fuse in the main consumer unit which connects up the cooker panel will be either a 30 or 45 A fuse, depending on the size of cooker to be fitted.

A wire is usually fed from the cooker panel and down the wall, this is chased into the plaster and emerges from the wall just about two feet from floor level. This is normally about the height at which the connecting terminals are mounted on the rear of the cooker. Due to the high current drawn by a cooker (6 to 12 kw) the terminals for the connections are usually in the form of large bolts locked in position by two brass nuts. The live and neutral terminals are usually mounted on ceramic insulators and the earth terminal is connected directly on to the case of the cooker. It is important to ensure that when the cable is connected to the cooker the terminals are tightened securely. Failure to do this could result in arcing and the terminals becoming burnt, giving rise to an ultimate short circuit and causing the fuse or circuit breaker to trip out.

A special terminal plate must be fitted to the cable where it emerges from the wall. This plate locks the cable securely, thus not allowing it to be pulled from beneath the plaster. A safety chain can also be fitted between the cooker and the wall; this needs to be of a shorter length than the connecting cable so as to take the strain should the cooker be pulled away from the wall for cleaning purposes.

How to replace a hot plate ring

Most modern cookers have a lift-up top plate to facilitate ease of cleaning. If this top plate is removed the hot plate and the connections will be seen and a simple change of element can then be effected. In some cases one lead from the hot plate needs to be connected through the on/off regulator switch normally mounted on the front panel. Should this be the case, providing the correct type of element has been purchased for the replacement, the connections will be quite self-evident. It is impossible to list all the numerous types of elements fitted as they vary greatly from manufacturer to manufacturer.

The oven elements will become visible for changing once the side of the oven has been removed, which is usually easily achieved by removing four screws—one at each corner of the plate. It cannot be stressed enough that the correct element must be purchased. **Do not try to fit a substitute, insist on the correct one for your model**. If difficulty is experienced in purchasing an element or hot plate ring consult your local Electricity Board office or write direct to the manufacturers, listing exactly what you require, and the make, serial number and model number of the cooker.

Fault Diagnosis

Cooker fails to heat up:

1. Check main fuse or circuit breaker.
2. Check connections to rear of cooker.
3. Check connections in cooker panel.

Hot Plate or Oven fails to heat up:

1. Remove hot plate and check with an ohm meter.
2. Remove switch and check with ohm meter—replace faulty part.

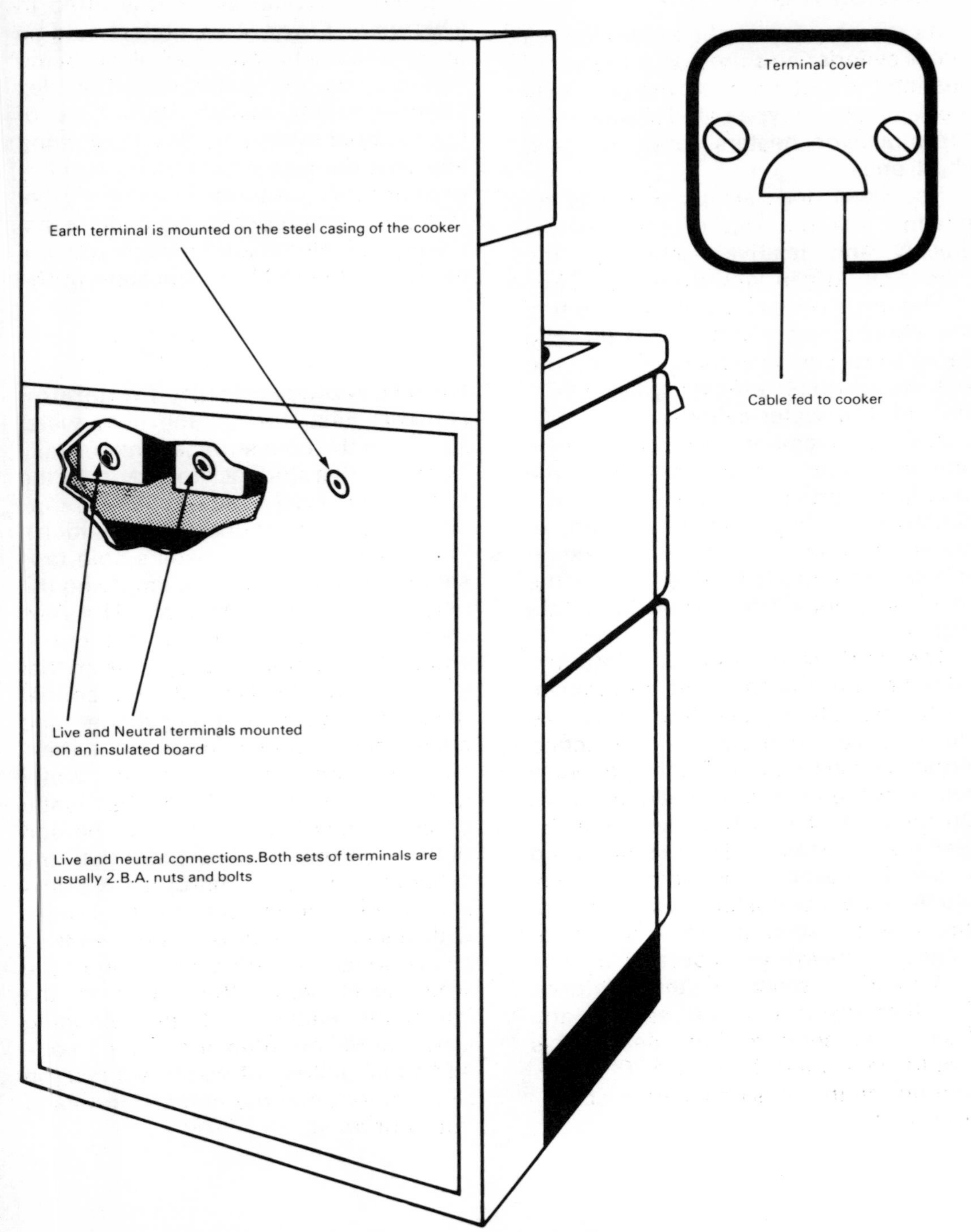

Cooker terminal plate and rear view of cooker showing location of terminal mounting bolts

Immersion Heaters

Immersion heaters are fitted into the hot water cylinder and provide a means of heating the water in the cylinder. There are two general types of elements used for immersion heaters, long ones and short ones.

The short ones are used mainly for heating just the top of the cylinder quickly and relatively cheaply; the power rating can vary from 1 to 2 kw.

The large ones are used for heating the whole cylinder and range from 2 to 3 kw. These can be expensive to run but will heat, quite efficiently, a 900 × 460 mm hot water cylinder.

A double element heater can now be purchased which incorporates both the long and short elements. This gives the consumer a choice on the amount of water he or she feels requires heating at any one time. It is ideal to use the small one, apart from when baths are required.

The immersion heater fits into the cylinder from the top, and on removal of the insulating jacket the protective plastic cover covering the connecting terminals will be seen. This is generally held in place by a large-headed screw. On removal of this box there will be seen the three terminal connections and a small plastic box some 25 mm square—the thermostat. In the centre of this will be seen a small dial with a central screwdriver slot, and on this is marked a series of temperatures. The thermostat can be set to any desired temperature but ideally this should be around 160° to 180° F, dependent on the outside weather conditions.

If the hot water cylinder is fitted in a bathroom the on/off switch should be fitted elsewhere, an ideal place being just outside the bathroom. The flex connecting the switch should be of butyl rubber three-core flex, preventing any heat damage which could result if ordinary PVC cable or flex were used.

A point worth bearing in mind is that if you fit an illuminated switch you will easily be able to check the state of the heater.

How to replace a faulty Immersion Heater. First turn off the cold water supply to the cold storage tank, usually by means of a stopcock found under the kitchen sink (in the older types of property this may be found in the garden). The upstairs and downstairs cold taps should then be turned off, emptying the cold pipe feeding the tank. The next step is turn on all hot taps in order to empty the water supply, apart from that left in the cylinder. Undo the top copper pipe connection to the cylinder and place a short length of hose into it, then, with the other end, suck the water through the tube. Once the water begins to flow make sure that the end of the hose is at a lower point than the cylinder—i.e. pass it through a window and hang it down outside the house. Only a small amount of water needs to be syphoned off, sufficient to bring the water level below the height of the immersion heater nut. If this syphoning is not carried out then about one to one and a half gallons of water will spill on to the floor once the immersion heater nut is undone.

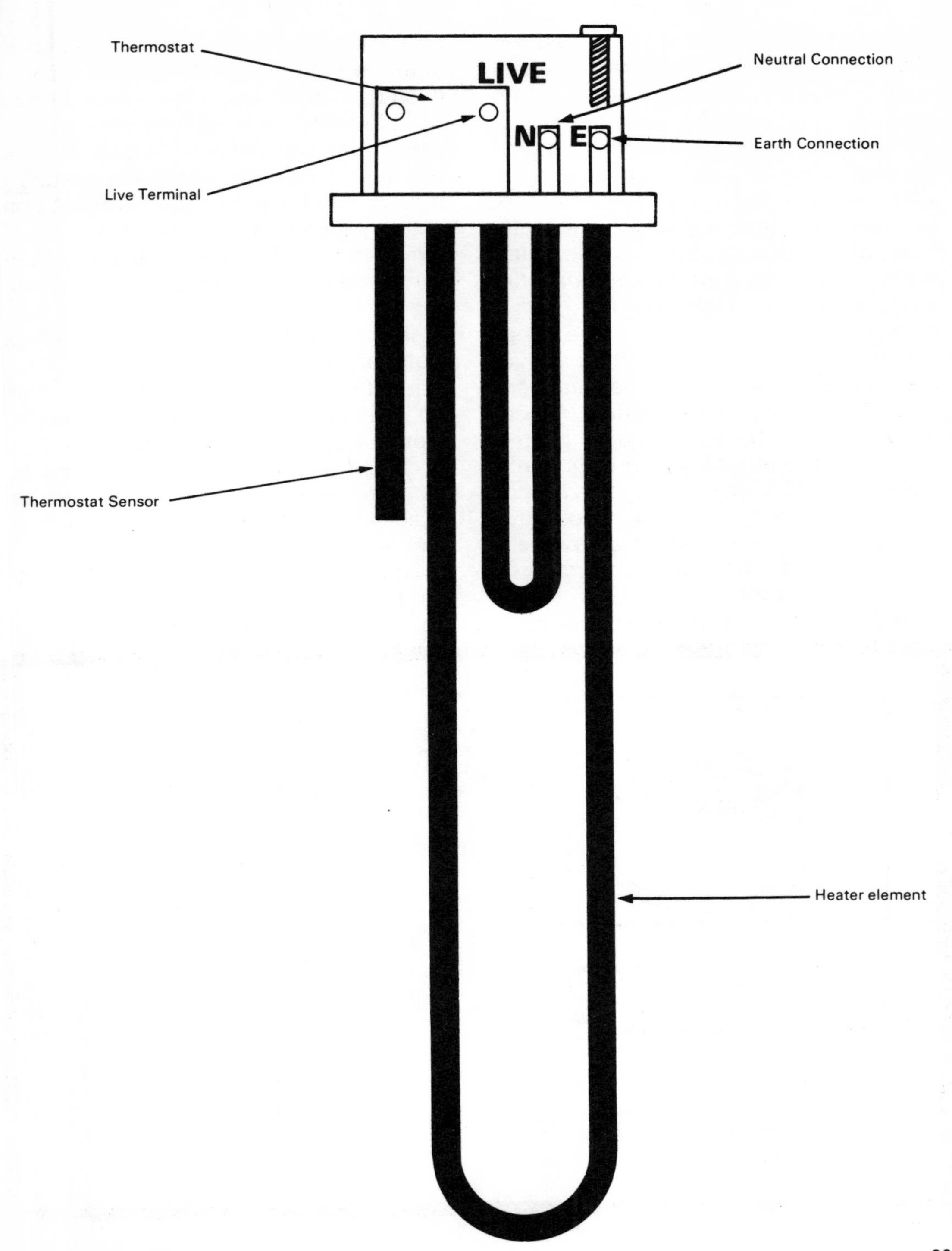

Inside an immersion heater

A very large spanner or a pair of 450 mm stilsons will be required to undo the nut (if you do not possess any then they can usually be hired quite cheaply from a tool hire firm or plumbers' merchants). Take great care when undoing the nut not to distort the cylinder as these cylinders are only made of thin copper sheet and are quite weak. Once the nut is undone the whole immersion heater can be withdrawn. If the element is of the long variety then the shelves, if fitted above the cylinder, will need to be removed in order to extract the element. Once this has been effected the new heater can be fitted. Before fitting the element, however, PTFE tape must be wound around the thread on the nut so as to effect a water-tight seal (this tape can be purchased from most plumbers' merchants). When this has been completed the heater can be fitted and tightened securely. The copper connections at the top of the cylinder can be reconnected once the heater has been fitted. Both hot and cold taps should then be closed and the cold water supply turned on; wait to check that both the disturbed connections are water-tight before leaving the cylinder. Finally, the terminations of the immersion heater can be reconnected and the thermostat set.

If only the thermostat on the immersion heater has broken, and this will be evident if either the water does not heat up at all or if it just boils, it can be removed and replaced by simply disconnecting the wires and withdrawing it. Once removed the thermostat can be checked by connecting this, together with a bulb and battery, into a circuit—it should be possible to switch the bulb on and off providing the thermostat is immersed in a glass of hot water.

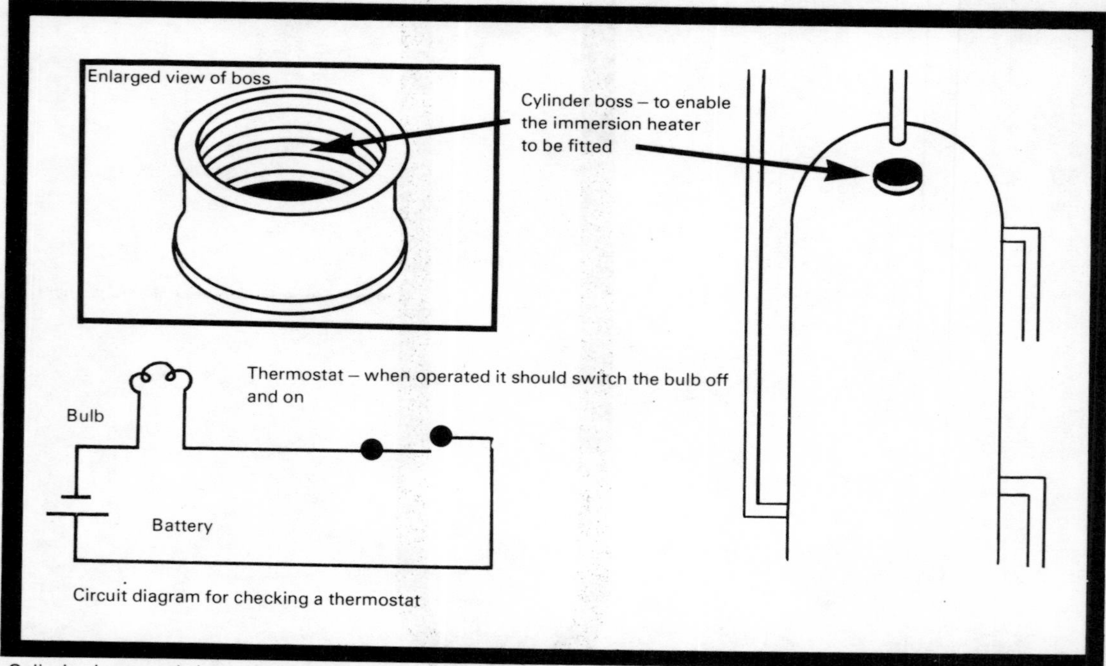

Cylinder boss and thermostat checking circuit

Table Lamps

Table lamps and standard lamps utilise the same type of fixings and connections.

How to replace the fitting at the top of a lamp. First remove the locking ring (A) to make removal of the shade possible, then separate the two halves of the connector by taking off the serrated locking flange (B). This will then expose the two leads connected into the spring loaded bayonets which in turn supply the current (amps) to the bulb. Once the two wires have been disconnected from the terminals the connector can then be unscrewed from the top of the lamp. The lamp holders can be purchased either with or without an on/off switch. There are only two connections in the lamp holder into which the live and neutral conductors should be connected. On older types of lamps it may be found that a brass connector is fitted; if this is the case it might be beneficial to refit the lamp with a plastic or bakelite one.

Replacement of flex fitted to a standard or table lamp is relatively simple providing that, before removing the old flex, a piece of string or the new flex is tied on to it; this eliminates the problem of trying to feed a piece of flex through what, in the case of a standard lamp, could be a $1\frac{1}{2}$ metre long hole. In some instances a clamping screw is fitted in the base of a lamp, this prevents the cable from accidentally being pulled out of the light fitting when replacing the plug on to the new flex.

Be sure to check that the fuse is of the correct rating.

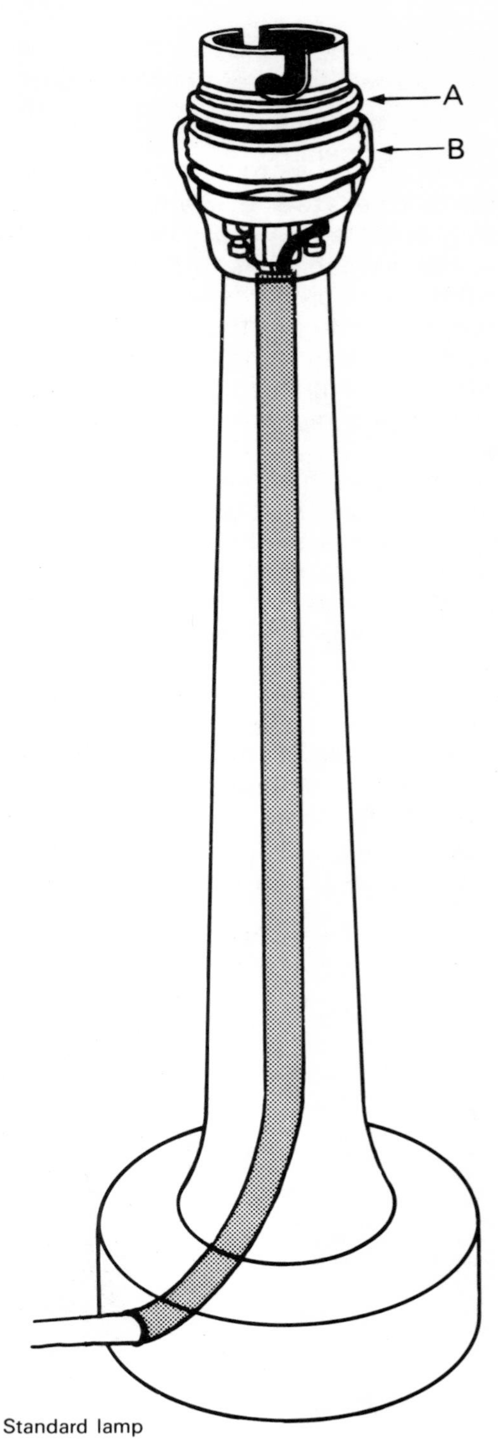

Standard lamp

Door Chimes

There are two types of door chimes:

1. Mains operated
2. Battery operated

Mains operated system. To fit a mains operated type of door chime a live lighting cable has to be found. To do this the floorboards will need to be lifted in a bedroom to find a piece of 3/029 or 1.5 mm² twin with earth cable, on to which will need to be connected a new piece of cable to fit the door chimes.

Removal of the floorboards is a relatively easy task. First go along the edge of both sides of the board with a sharp bolster chisel, removing the tongue from the floor board, then gently prize the board up. If some nails tend to be stubborn and hold the board, continue to exert a firm pressure upwards from under the board whilst hitting the top of it with a hammer. This will generally provide sufficient shock to release the board. Once the board is loose it will need to be cut into two above a joist—**do not saw it in half between two joists**. Once the board has been removed and the cable found the power must be turned off. The cable must be cut into and a new piece of cable connected on to it. The best method of doing this is to place a sharp knife in the centre of the cable and draw it along; the outer sheathing can then be peeled off for a distance of approximately 150 to 200 mm. The live and neutral conductors will be left exposed but still insulated.

The next step is to screw a 15 A joint box on to the side of a joist and lay the cable across it. Where each conductor rests on a terminal remove the insulation from the live and neutral conductors and place in the terminals, then strip back the insulation on the new piece of cable which is going to feed the door chime and connect this into the joint box. The lid on the box can then be secured.

The reason for stripping the existing cable in the way mentioned is to avoid having to connect three pieces of wire into each terminal, as this sometimes can prove almost impossible.

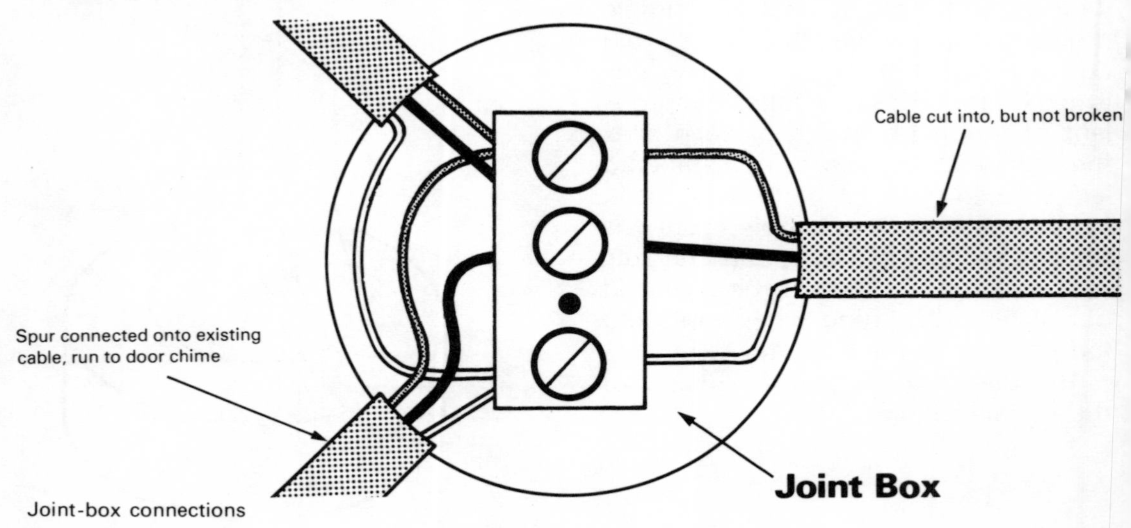

Joint-box connections

The new piece of cable which is to be fitted to the door chimes will be protected by the 5 A fuse in the consumer unit, in the same manner as the lighting circuit from which it was taken. This cable then needs to be connected into the chime transformer. A transformer is a device which can be made to increase the mains voltage from 240 V to, for instance, 500 V or reduce it from 240 V to, say, 6 V.

Transformers used for door chimes are usually built into a plastic box and have six terminals mounted into the plastic. Two of these are for the mains cable to be fitted and will be marked '240 V ac'. There is not normally provision for connecting the earth wire. **If there is no terminal leave the earth wire disconnected**. The other four terminals are used for connecting the single-strand two-core flex to the actual chimes. Normally one of the terminals will be marked 'common' and the other three may have numbers stamped by the side of them, and these numbers will indicate the output voltages between the common and that particular terminal. The normal value of output voltages is 6, 8 and 10 V. You will select the appropriate terminal at the transformer, depending upon the voltage at which the door-chimes work.

In general the transformer is mounted on the side of a joist between the ceiling and floorboards. The twin-core flex then needs to be fed to the push button and the chimes. Ideally, the floorboards that were lifted to install the transformer should be as near as possible above the door where the push button and chimes are to be fitted.

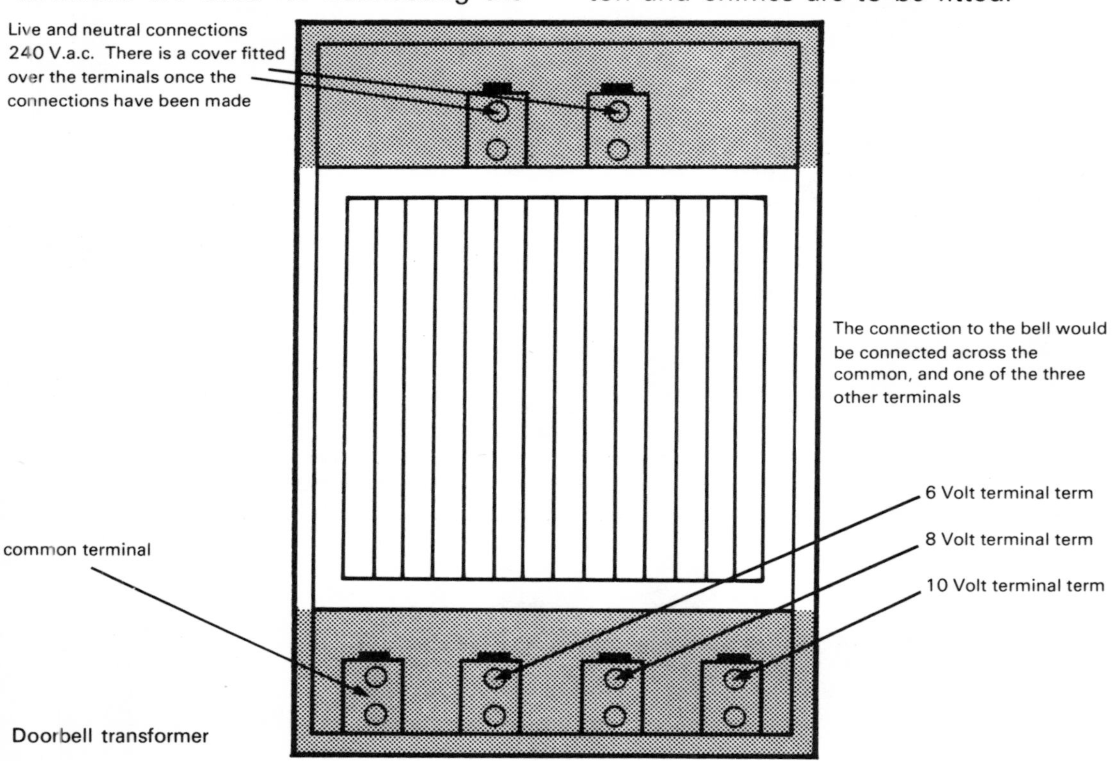
Live and neutral connections 240 V.a.c. There is a cover fitted over the terminals once the connections have been made
The connection to the bell would be connected across the common, and one of the three other terminals
6 Volt terminal term
8 Volt terminal term
10 Volt terminal term
common terminal

Doorbell transformer

To fit the actual box containing the chimes a masonry drill will be required in order to drill and plug the wall to provide a firm fixing. The twin-core flex must then be fed from the joist space down the wall to the chime box. To avoid damage occurring to decorations this cable can be surface-clipped to the wall and then chased into the plaster at a future date when redecorating is carried out. A wire must also be fed from the chime box to the push button switch which will be mounted on the outside of the door pillar. **Do not mount the switch on to the door** as continual opening and shutting will cause the flex to crack and ultimately break. With the transformer-operated door chimes an illuminated push button switch can be fitted which is useful in dark conditions, as it helps one to select the correct door keys, etc. This is not to be recommended for battery operated systems as the continual illumination of the light would flatten the batteries in a very short time.

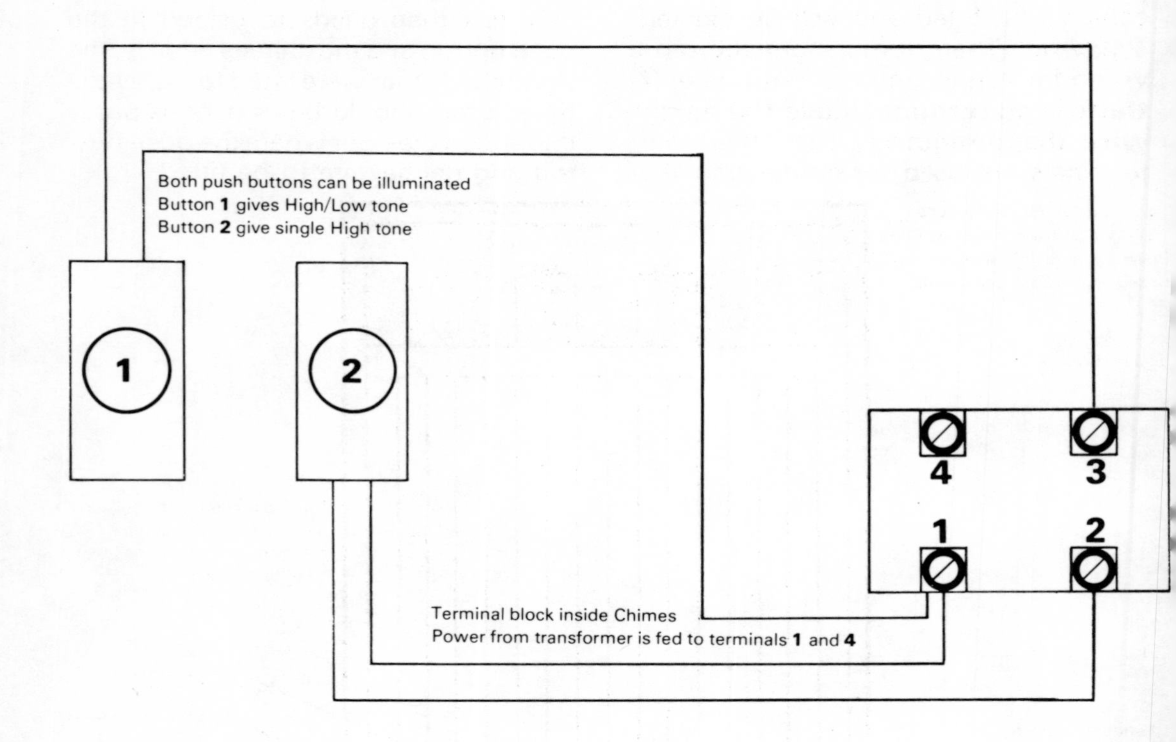

Doorbell connections for one- and two-button operation

Battery operated systems. The fitting for this type of system is exactly the same as for the mains operated system, apart from the fact that there are no connections into the mains supply and batteries are fitted into the chime box. Under normal conditions the batteries will last up to twelve months.

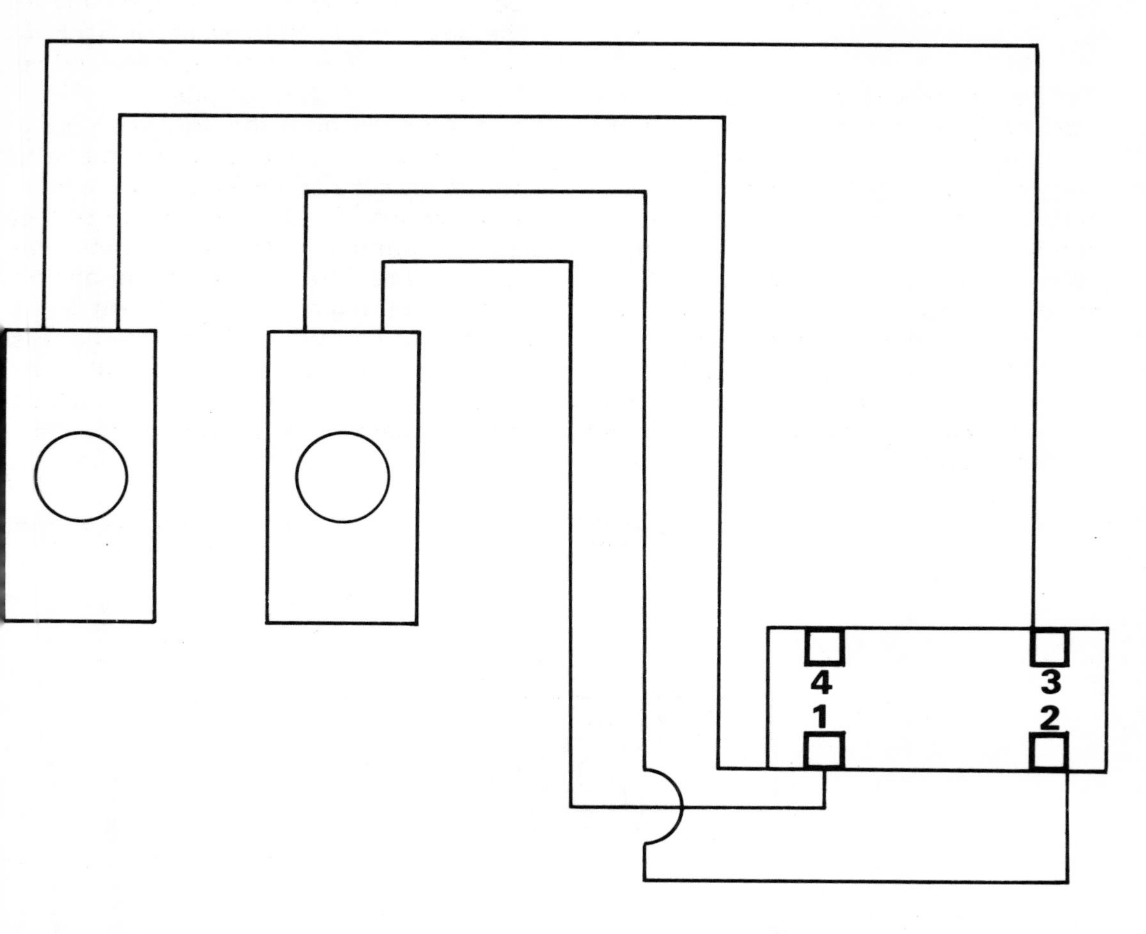

Connections for battery operated systems are the same as for a transformer one, with the exception of batteries fitted into the door chime case

Battery-operated doorbell system

Shaver Points

A shaver point can be fitted into a bathroom. It is the only type of socket that can and **under no circumstances must an ordinary 13 A socket be fitted**. Shaver points are only two-pin sockets and require no earthing through to the appliance, as electric shavers are double insulated and therefore require no earthing themselves.

To fit a shaver point. First of all the lighting cable must be located. If the bathroom is upstairs, which is normally the case, finding the lighting cable is relatively simple. If one enters the loft the lighting cable will be seen running in a continuous loop from light rose to light rose and ultimately back down the wall to the consumer unit. The shaver point can then be fitted as a spur to the lighting main or can be included into the ring main and form an integral part of the circuit.

When fitting a shaver point as a spur the ceiling rose cover in the bathroom must first be removed, and this will expose the cables feeding the light. In modern houses a 'three-plate rose' is used, which allows for a live and neutral conductor to be fed into the light rose and for a live and neutral conductor to be fed out of the rose and ultimately to the next rose in the circuit, or back to the consumer unit if that particular rose happens to be the last one on the circuit. Also fed from the light rose are a live and neutral which connect the rose to—in the case of the bathroom—a pull-cord switch. The live conductor of this cable is connected to the two other live conductors. The neutral is connected into the third terminal in the rose on its own. The flex which feeds the light is connected between the neutral conductor and the other terminal, which has connected to it two neutral conductors.

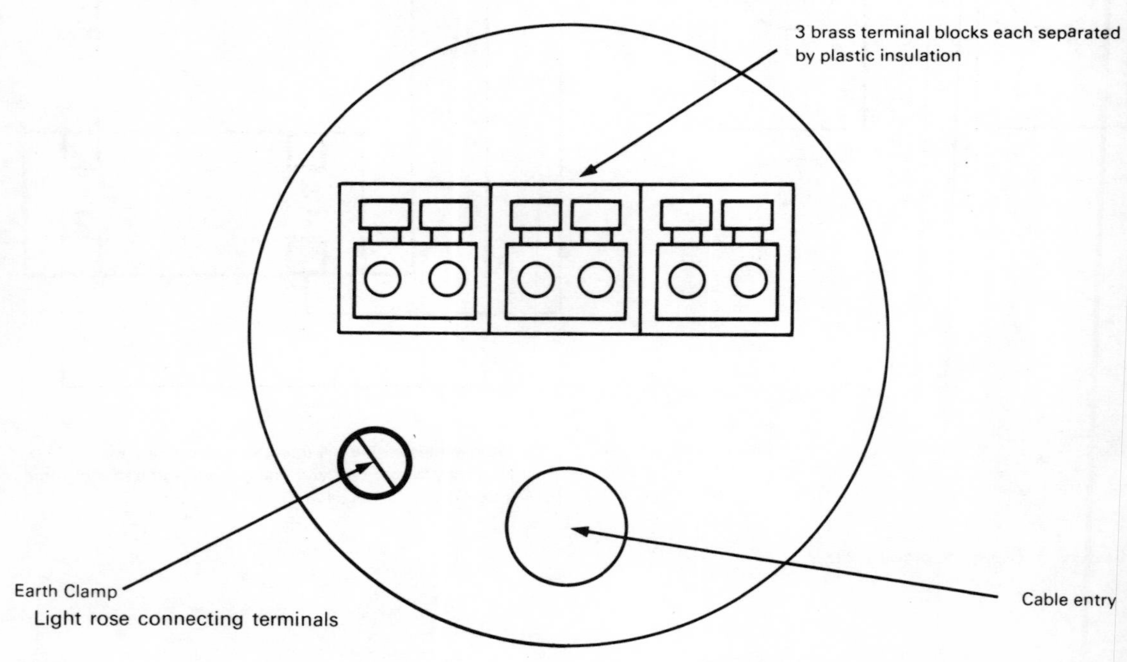

Light rose connecting terminals

To fit the shaver point as a spur a live and neutral conductor must be connected into the rose on to the two live and neutral terminals. This cable is then fed to the position at which the shaver point is to be fitted. One point which must be noted is that, although there has been no mention of the earth terminal, it must be connected and in all lighting fittings, i.e. roses, switch boxes etc., a small brass terminal will be found. This brass terminal is for the connection of the earth lead. The purpose for this earth terminal is so that a continual earth loop is fed round the house, and should, for instance, a steel light fitting be fitted, it would automatically be earthed.

Shaver points usually incorporate a strip light and are best fitted above a mirror. The light can be one of two types—incandescent or fluorescent. The fluorescent ones probably give a better shadow-free, brighter light but in some cases they can tend to make a buzzing noise. The incandescent type are made in the same way as a normal light bulb but are encapsulated in a long cylindrical glass envelope, instead of the normal pear shape. A point in favour of this latter type is that it lights instantaneously, whereas the fluorescent ones can take seconds to ignite. The shaver point is normally mounted in the end of the light fitting.

To fit the shaver point into the ring main to make it an integral part of the circuit, one of the cables feeding a light rose must be removed and fed down to the position where the shaver point is to be fitted. If the cable is not long enough then a 15 A joint box must be fitted and another piece of cable connected on to it. When this has been fed to the light a return piece of cable must be fitted and fed back into the light rose to the same point from which the other piece was removed. This method is the one to be preferred as it does mean that the ring is continuous and that a spur has not been fitted. However, this method tends to be more expensive than the one mentioned previously.

Another method of connecting an appliance or fitting into a ring main is to cut one piece of 3/029 or 1.5 mm^2 cable and on to each end connect a 15 amp joint box, and then from each joint box two separate cables should be run down to the shaver point. Remember to connect the earth terminals.

Actual maintenance of a shaver point is not necessary. The only thing that could possibly occur is for the lighting

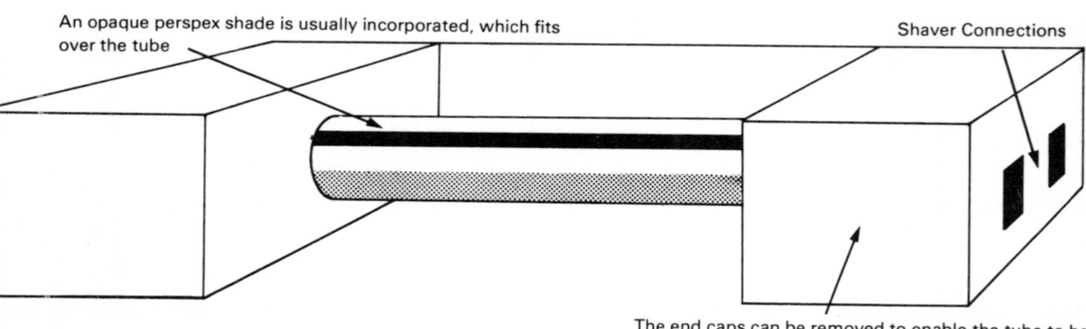
An opaque perspex shade is usually incorporated, which fits over the tube
Shaver Connections
The end caps can be removed to enable the tube to be replaced

luminated shaver point

tube to blow or the terminals, which connect the shaver, to become coated in verdigris. The latter would really only occur in extremely damp or steamy bathrooms, and then could take some considerable time to do so. If this should happen the light fitting should be removed and the terminal cleaned with a mild abrasive (emery cloth) or, in less severe cases, metal polish should be sufficient.

Bathroom Heaters

There are basically two types of heater:

1. Circular type fitted in place of the bathroom light (this type of heater usually incorporates a light also).
2. Strip wall heater.

Circular heater. To fit this type of heater the light rose must first be removed to expose the two cables forming the feed in and out of the rose and a further piece of cable which connects the switch to the rose. If the circular heater and light do not have a three-plate fixing inside (see previous section on shaver points) connections must be made using a barrier strip connector.

The next step is to connect the live, neutral and earth conductors into the barrier strip and then connect the switch wire between the live terminal and a spare terminal in the barrier strip. The wires from both the neutral terminals are then connected and these should be fed to the heater. The heater element is generally only a 750 W element, unlike normal electric fire elements, and is encapsulated in a silica glass tube. This is to prevent steam from coming into contact with the element which could cause a short circuit. The heater would normally be switched on by a pull-cord hanging from the heater.

Strip wall heater. This type of heater can either be fitted into the ring or run from a spur, and the method of connecting this type of heater is as described under shaver points (see page 46).

The wall-mounted type of heaters do not have a light incorporated in them. If the element should fail to heat up, in

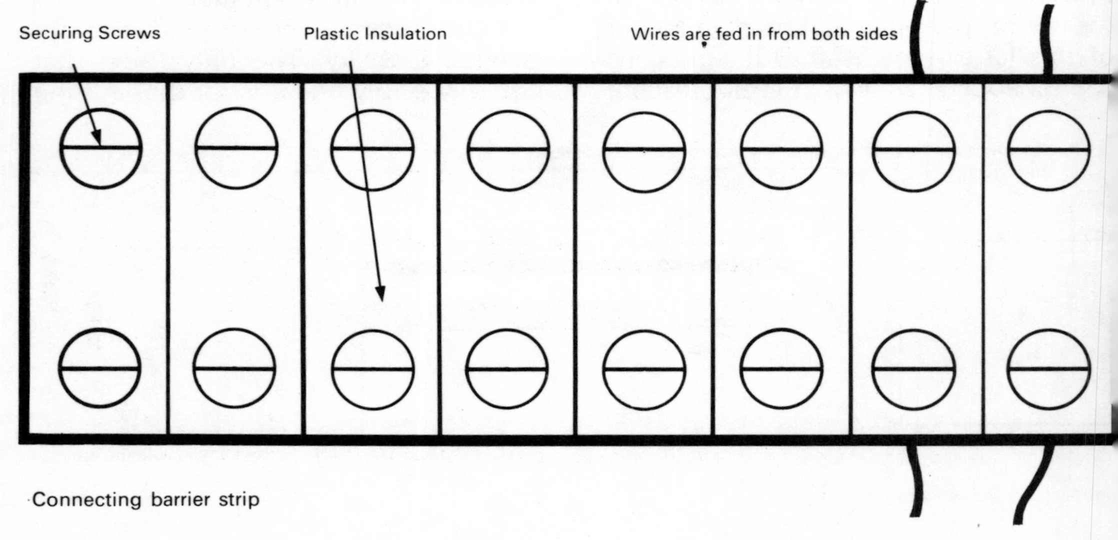

Connecting barrier strip

either case, this can easily be removed by undoing the two securing bolts at either end of the element in order to check it with an ohm meter.

Electric Irons

An electric iron basically consists of an electrical heating element encapsulated in a steel or aluminium plate forming the base of the iron. The temperature of the iron is governed by a thermostat which is adjusted by the knob on the top of the iron. The knob is usually graduated by the words nylon, rayon, cotton, etc. Generally, electric irons will work satisfactorily for a great many years, but should they malfunction the common faults are either a faulty thermostat or a broken element.

If the iron breaks down, the element and the thermostat should be checked. This is achieved by first removing the heat control knob which will allow the thin steel plate immediately under the knob to be pulled off. This will expose a large hexagon-headed nut—this nut must also be removed. The heel (back) of the iron then has to have its back plate removed so as to expose two terminals, these are live and neutral. The terminal screws have to be disconnected so as to allow the element and thermostat to be removed from the casing. The live terminal will be found to run to the thermostat and then on to the element, whilst the neutral is connected directly to the element. The thermostat can easily be detached by undoing the one securing screw. If the contact points are badly pitted it is advisable to replace the thermostat. The element will need to be checked on an ohm meter and should you not possess one then most electrical retailers will check the element for you free of charge. If the element is found to be faulty then, of course, this must be replaced. Assembly of the new element and thermostat is exactly the reverse procedure of the disassembly.

Vacuum Cleaners

A vacuum cleaner consists of an electric motor to which is connected a fan. This fan causes a vacuum resulting in a sucking action which pulls the dirt and dust from the carpet and into a disposable paper bag (in the older models this is a non-disposable canvas bag).

There are two types of vacuum cleaner:

1. *The upright type*—these have fitted to them a revolving brush and beater which it is claimed helps remove the dirt and lift the pile.
2. *The cylinder type*—these have a long hose attached to them on to which are connected the various cleaning tools.

Both types of cleaner work on almost the same principle and that is of a vacuum being caused by an electrically powered fan. Therefore, the only electrical fault that is likely to occur is a motor failure, apart from obvious cable fractures or loose terminals.

There are two carbon brushes fitted to the motor and should the motor fail it is generally due to the brushes becoming worn. If the motor is visually examined two plastic serrated caps approximately 15 mm high by 15 mm diameter will be seen. If these are unscrewed they will spring away from the motor as under these plastic caps are fitted springs on to which are connected the carbon brushes. These brushes transmit the power from the mains input on to the commutator (the revolving part of the motor). If after replacing the two carbon brushes the motor still fails to run the fault will either be in the commutator section or the outer set of windings in the case of the motor. If the latter is the case then the windings will need to be checked on a special, rather expensive, piece of equipment, by a firm specialising in motor rewinds.

Hair Driers

Hair driers consist of a heater element and an electric motor which drives a fan. The heating element is usually made in the form of a loose wound spring which is wrapped round a former, the former being made of a heat-resisting insulating material, usually ceramic. The heater element is normally fitted with a thermostat to protect the element from overheating. The thermostat, which is not adjustable and is set by the manufacturers to cut out at a predetermined temperature, forms an integral part of the heating element. The fan is driven by an electric motor which is generally a synchronous motor that requires no brushes and virtually no maintenance.

If a fault occurs in the hair drier, the fault can be isolated to either a motor failure or a heater failure. If the motor is found to be faulty then the probable cause of this will be a burnt winding. This can sometimes be seen but, if the windings are burnt on the inside, then no visible signs will show. The only remedy for this type of fault is either to replace the motor or to take it to a motor rewind firm in order to get the winding rewound. It must be pointed out, however, that the latter method could prove to be more expensive than replacement of the complete motor.

If the heater element or themostat are faulty then these also must be replaced. When purchasing a new heating element it may be quicker to write direct to the manufacturers and order a replacement direct from them. To expose the element and thermostat it will be found that if the five or six case securing screws located in the side of the case are removed, the case will split into two leaving all the electrical parts easily accessible.

The element will have a live and neutral wire connected to it. The two screws retaining the live and neutral conductors must be undone. If there are any screws securing the element to one side of the case, these must also be removed to allow the element to be withdrawn and the replacement fitted.

In the case of the motor, it will be found that there are two conductors, one live and one neutral, feeding the motor. These must be disconnected and any securing screws holding the motor in place must also be removed to allow the motor to be withdrawn and the replacement fitted.

How to Change a Central Heating Pump

A great majority of houses have some form of central heating system fitted. At least 70 per cent of these systems rely on radiator heating and this method requires a pump to circulate the hot water round the system. The pump is mounted in a steel casing which incorporates the motor armature. Fitted to this armature is an impeller which is a steel disc with raised veins which, when turned, circulates the water round the system. This part of the pump is similar to a water pump fitted in a motor car.

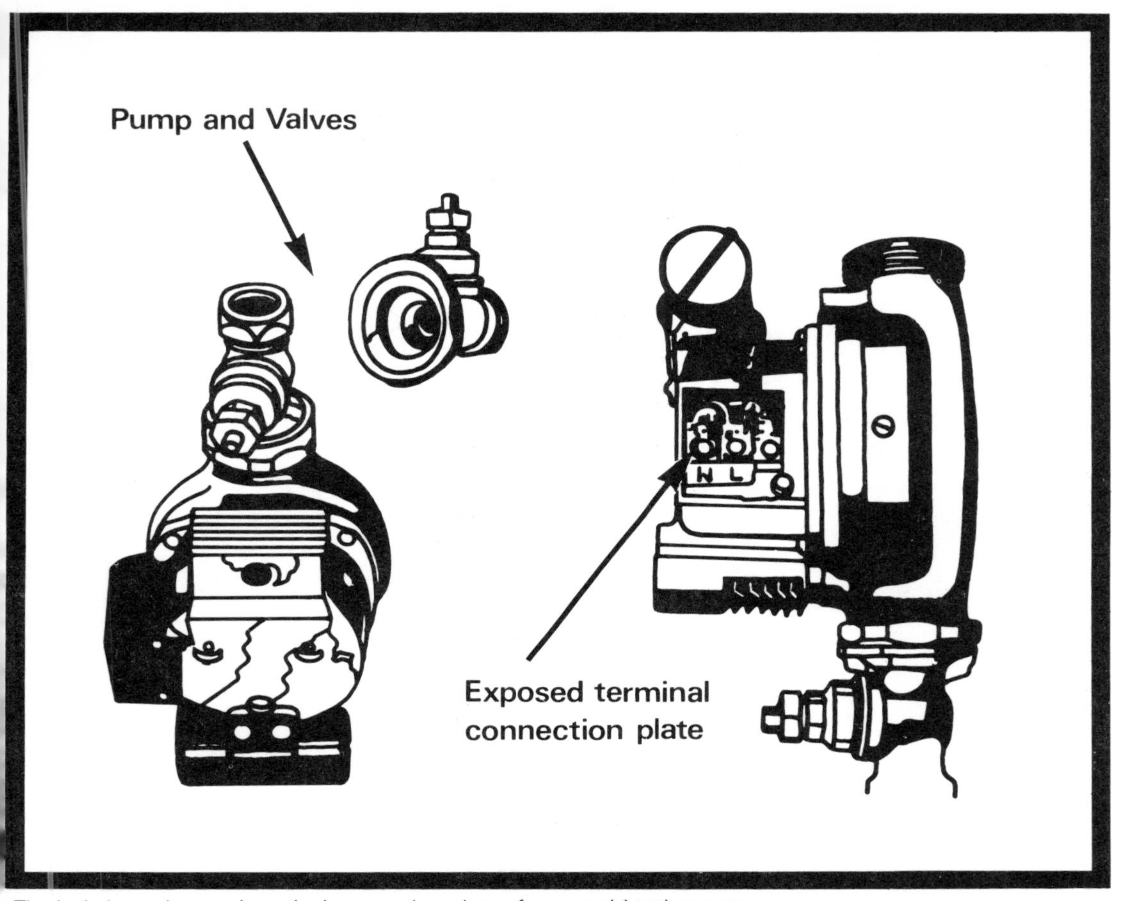

The isolating valves and terminal connecting plate of a central heating pump

Central heating pumps are usually fitted with a variable speed switch, normally graduated by the numbers one to five. If fitted in a two-storey house the pump will only need to be set on numbers two or three but, if installed in a three-storey house, then the pump may require to be set as high as number five. However, should the pump be set too high it will pump the water with such a force that the water will be ejected from the expansion pipe into the heating supply tank fitted beneath. This tank is usually located in the loft space.

Should the pump malfunction in any way, such as seizing or making a noise due to worn bearings, then it will need to be changed. To change the pump, two valves, which will be found one on either side of the pump, have to be turned off in a clockwise direction; this will isolate the water to the pump. Two large nuts connecting the pump to the copper tube then have to be undone and this will allow the pump to be withdrawn from the piping. Once this has been carried out the electrical connections can then be removed in one of two ways. The most common method is simply to withdraw the plug from the pump thus enabling it to be

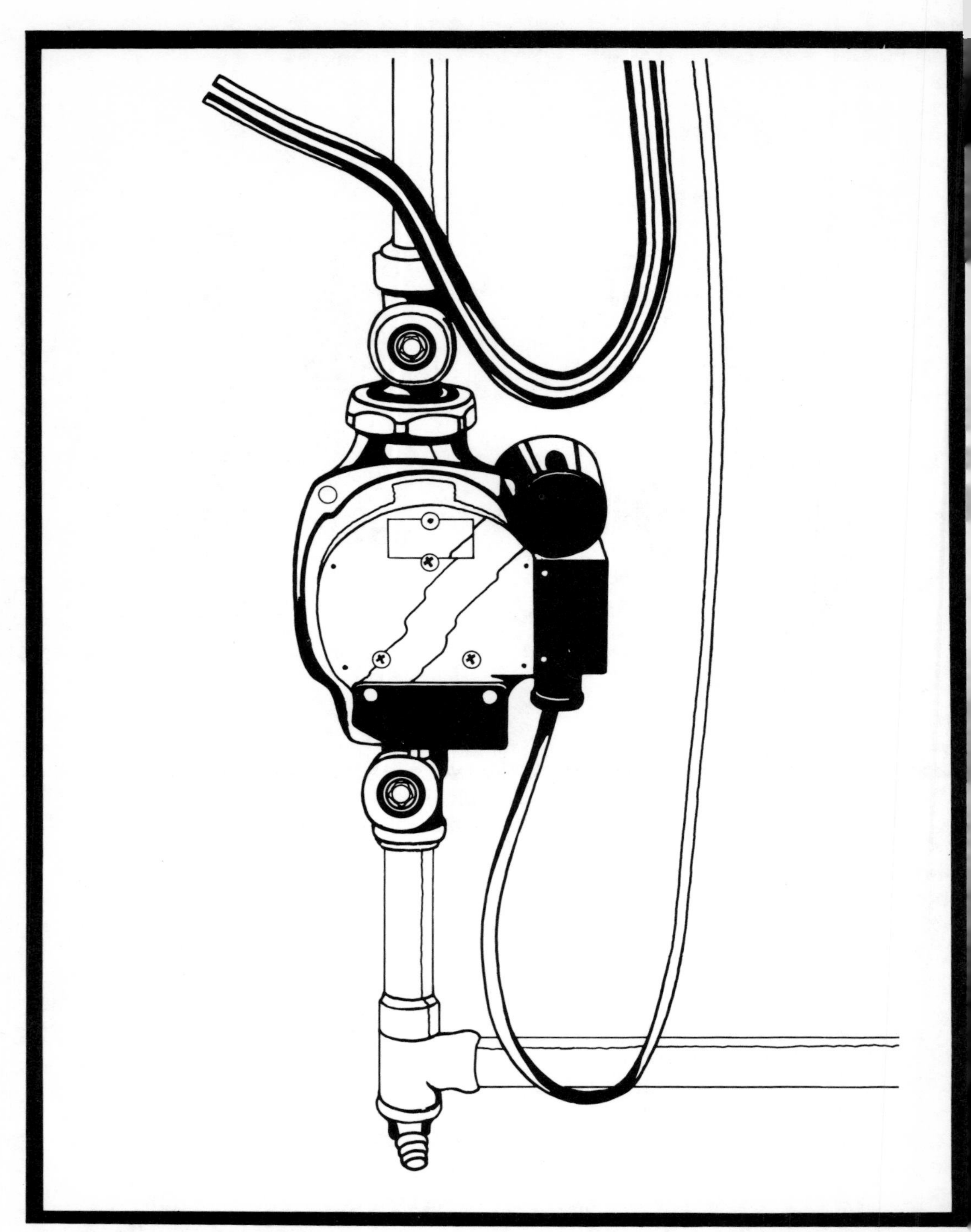

Central heating pump in position

completely removed. In older types of systems it may be necessary to remove an inspection cover and disconnect the input conductors from the motor terminals.

Replacement of the new pump is exactly the reverse process of the disconnection.

To ensure that the valves are completely watertight foliac pipe compound or PTFE pipe tape can be added to the joints before reassembly. The isolating valves should be reopened and any air that may be in the pump impeller will then be allowed to flow into one of the radiators, which will then necessitate the bleeding of each radiator. Following this the electrical section must then be replugged in or the terminations connected on to their respective terminals in the pump.

Once the pump has been replaced and the electrical and plumbing connections have been satisfactorily completed, then the new pump must be set to its correct setting. To do this a screwdriver will need to be inserted into the screw slot, pressed inwards and then turned to the correct setting.

Generally, central heating pumps can be purchased on an exchange basis, and therefore when buying a new pump it is advisable to enquire if an allowance will be made on the old pump should it be returned once the new one has been installed.

Rewiring and Extending Existing Ring Mains

Wiring a Cooker Panel

To wire a cooker point from the mains input to the actual cooker the first item to check is the type of cooker to be installed and whether it will require a 30 A or 45 A fuse. Most cookers will only require a 30 A fuse. The consumer unit should be checked to see if there is a spare fuse space. If there is, the cable can be connected into it but, if not, then a new fuse box must be purchased. A single-way type would be satisfactory but it is advisable to fit a two- or three-way, thus leaving one or two spare fuse spaces should other extensions or additions be required in the future.

If a new fuse box has to be fitted then the live feed to the new consumer unit

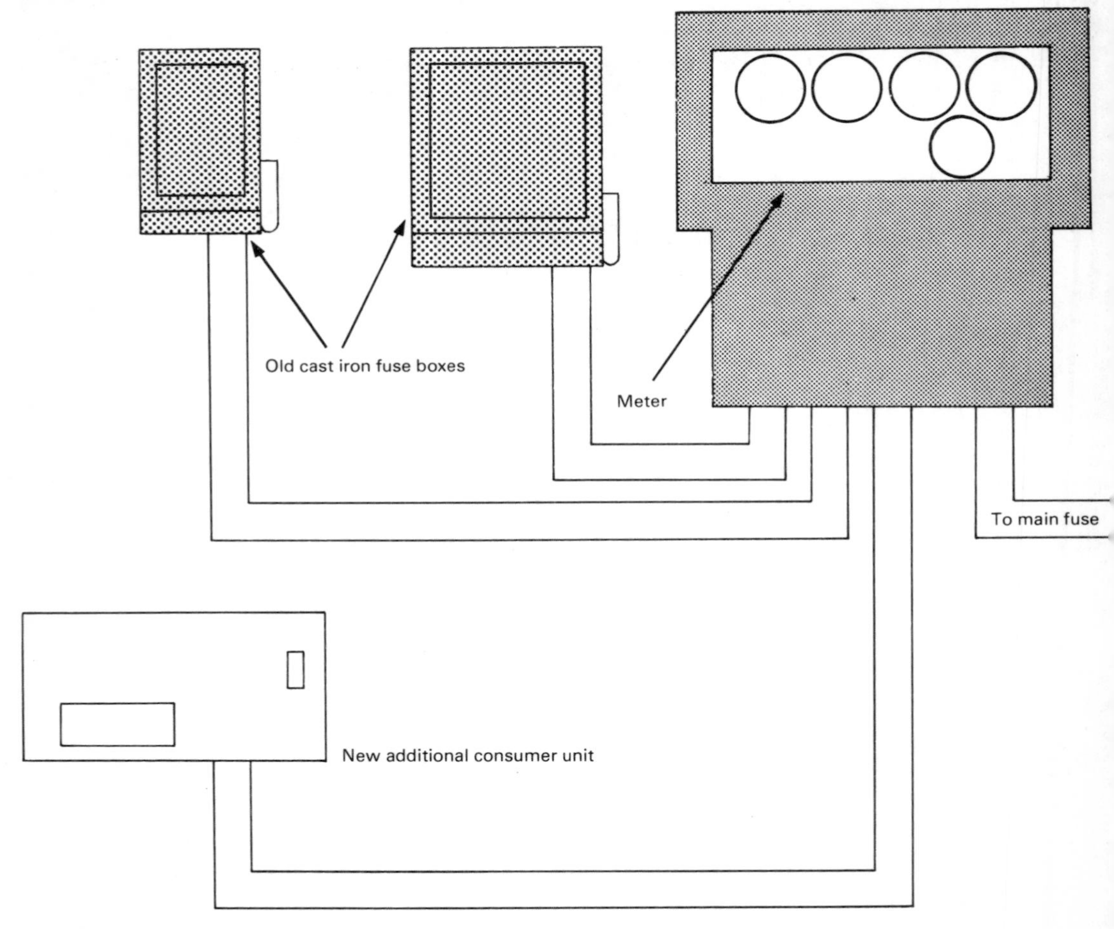

Old and new consumer units connected to electricity meter

must be fed from the meter. To do this the Electricity Board **must** be notified of what you intend to do in order that they can make arrangements to check the completed work. The lead seals can then be removed from both the electricity meter and the main 60 A or 100 A input fuse, and then this fuse can be removed. This will isolate the complete electrical installation in the house including the meter and the consumer unit already fitted. The front cap of the meter can next be removed and connected to a new piece of cable (meter tails), which has to be 16 mm^2 double insulated cable, and this then has to be fitted into the new consumer unit. Following this the meter cover can be refitted and the main fuse replaced. Power to the house will now be reconnected and a live and neutral feed will be made to the new consumer unit.

In the new unit there will be found a large brass terminal block for the earth connections, from this terminal a piece of single insulated green and yellow striped insulated 6 mm^2 cable should be bonded (connected) to both the gas and water pipes, notably on the supply side of the stop valve and gas valve. Another piece of this cable should then be connected to the steel casing of the mains input cable. If, however, a terminal is not fitted to this cable the Electricity Board will, if notified, fit one for a small fee.

The new consumer unit is now ready to receive any new appliance which is going to be connected to it.

If a number of small fuse boxes are already fitted it may be preferable to scrap these and fit a new, larger consumer unit to replace them, and this will also provide the required number of new fuse spaces for the intended additions to the circuit. Whichever method is adopted it must be realised that in order to render the consumer unit terminals and the meter dead (switched off) **the main sealed Electricity Board's fuse must be removed in order to render the circuit safe**. Do not attempt to connect a new consumer unit without first switching off the power—people have been known to do this in order to avoid calling in the Electricity Board to check the work once it has been completed. Once the new consumer unit has been fitted work can commence on fitting the new cooker panel.

There are a number of different cable sizes which can be used, depending on the size of the cooker and the distance between the cooker panel and the consumer unit, but 6 mm^2 double insulated

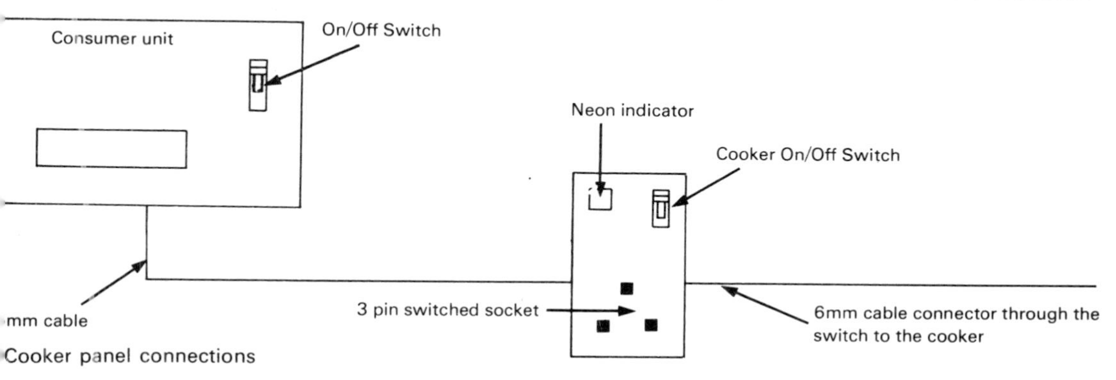

Cooker panel connections

cable with earth will be of sufficient capability to run all types of domestic cookers.

If the house is fitted with wooden floorboards it is probably best to run the cable under the floor to the kitchen, or wherever the cooker panel is required. If the floors are made of concrete then the cable will need to be routed up the wall, under the upstairs floorboards and back down to where the cooker panel is to be fitted. It must be noted, however, that in fitting a cooker panel, unlike a normal socket, it does not have to be on a ring, i.e. there does not have to be a return wire fed back to the consumer unit.

The cooker panel usually incorporates a 13 A socket and can be made with small neon lights to give an immediate visual check to see whether it is switched on or off. When actually mounting the cooker panel on to the wall be sure that the terminals are extremely well tightened as the cable will be carrying a load of anything up to 7–10 kw.

From the cooker panel a piece of cable must be fed to the cooker. Normally this is chased into the wall below the plaster and emerges from the wall at approximately the same level as the terminals on the cooker. At this point a terminal plate must be fitted to prevent the cooker from being pulled away from the wall which would ultimately pull the cable out of the plaster.

Installing an Immersion Heater

Again, the first step to be taken is to see if there is the facility for connecting a spare fuse in the consumer unit; if there is not a new fuse box and consumer unit must be fitted. An alternative to this is to replace the old consumer unit(s) with a larger new consumer unit, capable of accommodating the existing fuses and having sufficient space left to fit more as and when required. You must remember to use 16 mm² double insulated cable for the meter tails when connecting the consumer unit to the electricity meter. In most cases the immersion heater will need a 15 A fuse but, in a small number of cases, a 20 A fuse may be required. In order to check if a 15 A fuse is sufficient the wattage rating of the heating element must be found (this is usually stamped on the top of the element). The fuse value can be calculated by working out the following formula

$$A = \frac{W}{V}$$

An example of using this formula can be found on page 13.

The cable will need to be of 2.5 mm² twin with earth. Only one piece of cable will need to be fed to the immersion heater and no return will be necessary. It may take some considerable time to feed the cable from the consumer unit to the copper cylinder, as it may mean it traversing from ground to upstairs level and also from one side of the house to the other.

It is important to remember when chasing cables into plaster that they should always be run vertically. You will then be able to guess where the cable runs and avoid it if ever you need to drive nails or picture hooks into the wall.

When running the cable between the joist space it will be necessary either to drill holes in the joist or to cut small chases in the top of the joists. If holes are to be drilled then they should be 50 mm from the top of the joist, this

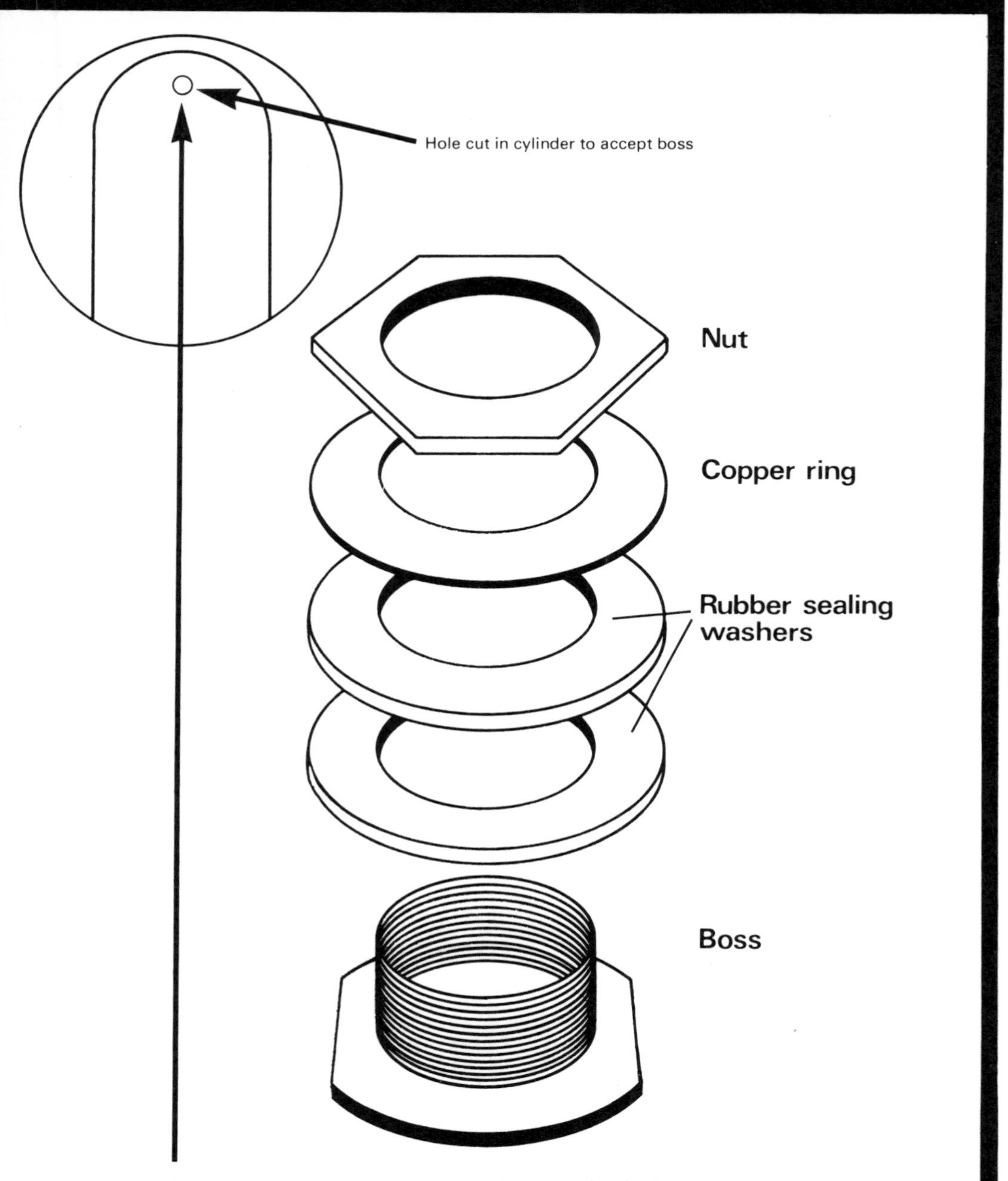

xposed view of cylinder flange

again will prevent the cables from inadvertently becoming nailed. If chases are to be cut then they should be in the centre of the joist in order to avoid the cable being damaged when the floorboards are renailed.

Once the cable has been fed through to the cylinder, which is normally in the bathroom, a switch must be fitted. If the cylinder is in the bathroom then the switch must be fitted outside the room to eliminate the possibility of steam causing a short circuit in the switch. From the switch another piece of cable must be run to the cylinder. It is advisable to use butyl rubber 3-core flex for this purpose as it is heat-resistant and will endure the continual heat applied to it from the cylinder.

If the cylinder has an immersion heater boss (usually a 75 mm diameter hole with an internal thread) this is normally blanked off with a large brass nut. Once the cylinder has been drained (as mentioned on page 38) then the boss can be removed and the immersion heater fitted. If, however, the cylinder has not got a boss, or if the cylinder is fitted close to the ceiling, thus making it impossible to fit the element into the cylinder, then a special adapting thread ring must be fitted. The adapting ring can be purchased from a plumbers' merchants. To fit the adapting ring a hole will have to be cut in the cylinder, which can easily be achieved by drilling a series of holes close together in a circle of 75 mm diameter. Once the holes have been drilled they can be cut into one large hole by using a pair of tin snips, taking great care not to let the waste copper drop into the bottom of the cylinder. With a half-round file the hole can be filed out until the boss, when tried upside down, will fit tightly into the hole. It will be noted that the boss will have a flange approximately 20 mm wide round the base. This is circular with the exception of two flats filed on it—these two flats allow for the boss to be fitted through the hole in the cylinder.

Once the hole has been adjusted and PTFE tape wrapped round the threads of the boss, the boss can be fitted. In order to do this the boss must be placed on its side and fed into the cylinder, then by carefully tipping the boss in a horizontal position it can be pulled upwards into the hole. With great care remove one hand at a time so that the locking nut can be placed on to the top of the thread bush. With one hand hold the boss and with the other hand hold the nut and turn until it is tightened. Finish this off securely with a spanner or pair of stilsons. Once this work is completed the immersion heater can be fitted (as described on pages 38–40).

Lighting Circuits

If an extension to a lighting circuit is required, it must be decided how many extra lighting points are going to be needed. If it is only two or three then the existing lighting ring can be con-

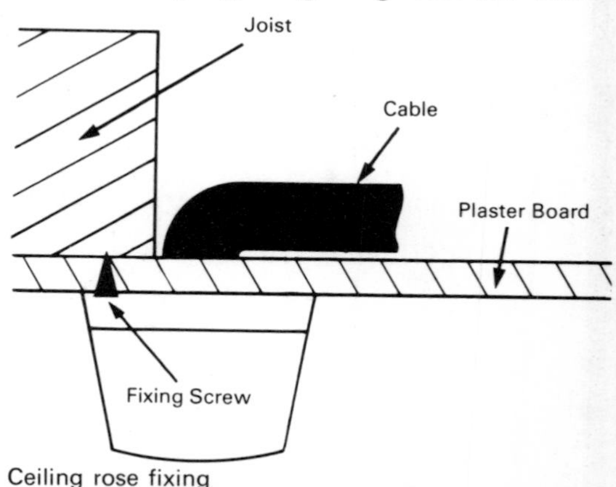

Ceiling rose fixing

nected into, provided no greater than eight lighting points in all are connected. Connecting into the ring can be achieved in two ways, as described on pages 46–8, but probably the easiest method is to locate a piece of 1.5 mm² cable, cut it in two and connect two 15 A joint boxes—one to either end. From one joint box a new piece of 1.5 mm² cable must be run to the position where the first light is to be fitted. The light rose can then be fitted to the ceiling, preferably on to a joist. When fitting the rose to the joist it should be put to one side of the joist, which will allow the cable to be run down the side of the joist and into the rose.

Another piece of cable then has to be taken to the newly installed light rose, and this should be run to the next position where a light is required. Repeat this process until all the new positions have been fed with an input cable and an output cable. From the last lighting rose in the new additions feed the outgoing cable back to the second of the two 15 A joint boxes; this will ensure that all the lighting roses are on the ring. The two pieces of cable protruding from each light rose can then be connected. The two live conductors have to be connected to one of the terminal blocks and to one other terminal block the neutral conductors. This will leave one spare block of terminals.

Once the above work has been completed the switch wire will need to be connected. First decide where the switches are to be fitted and fix a plaster-depth steel box into the plaster, securing it either by cement or, preferably, a wall plug and screw. If, as normal, the light switch is near to a door the cable can be fed up the wall behind the architrave on the door casing. Normally there is a reasonable gap suitable to feed the cable in once the architrave has been removed. This will conceal the switch wire up to the height of the door. From this point on it is a matter of personal choice whether to surface fix the cable (clip to the surface of the wall) or to chase a groove in the plaster and bury it. The latter results in a more professional looking job, but some people tend to surface fix until they next redecorate the room and then chase in the cable at that stage. The switch must be fed to the light rose, and the live and neutral conductors must then be connected. The live has to be fitted into the same block of connectors as the other two live conductors. The neutral is connected into the third, as yet, unused terminal block. The lighting flex then has to be run from the two neutral conductors to the lamp-holder. The final connection and probably, from the safety factor, the most important is to connect the three single-strand earth wires to the small terminal in the light rose. The reason for wiring in this fashion is to eliminate having a live wire permanently connected to the lamp-holder, this method of connection being known as a 'switched live'. In some older types of houses the neutral wire was the one which was 'switched', resulting in a permanent live terminal at the lamp-holder.

If more than three extra lights are required it may be more advantageous to fit a new consumer unit or use, if available, a spare fuse in the original consumer unit. If a new lighting ring is installed it is important to ensure that the cable is fed from the consumer unit to the lighting points and then back to the consumer unit. The amount of lights permitted on a ring is not clearly defined. There have been cases where only 1.0 mm^2 cable has been used and one ring has had all the upstairs and downstairs lights connected to it, but I would suggest that 1.5 mm^2 is always used for lighting and not more than eight lights are run from one particular ring.

Normally it is better to have one lighting circuit for upstairs and one for downstairs. Wall lights, if fitted, can be fitted on a separate fuse. In the event of a power failure of the lighting circuit, by fitting the wall lights on a separate fuse, it will at least ensure that a light of some description is available whilst the other ones are being repaired.

If a garage light is required and the garage is an integral part of the house, then this light can be included in the downstairs circuit. If, however, the garage is some distance from the house then a special fuse must be added in order to run a cable from the consumer unit to the garage (this problem is dealt with on page 63).

One advantage of fitting a new consumer unit and completely new circuit is that the installer knows, for future reference, exactly where the circuit runs and what is connected to it. If, however, there are a multitude of differing types of consumer units then it is preferable to disconnect the lot completely and purchase a new consumer unit large enough to take all the existing circuits plus any new additions and, if possible, leaving one or two spaces for any additions that might be required in the future.

Power Circuits

13 A power circuits can be fitted anywhere throughout the house and an unlimited amount of sockets can be fitted if the floor space is less than 100 square metres (1000 square feet), which covers most normal three and four bedroomed houses. In general, although there is no limit to the number of sockets that can be fitted into a ring, it is preferable to fit one ring upstairs and one downstairs. From a 13 A ring main one spur is allowed and not more than two sockets can be fitted on to this spur or one permanent corded outlet device (e.g. oil-filled electric radiator, central heating boiler point). It is possible to connect on to the existing power circuit by using the method mentioned previously for lighting, with the exception that the joint boxes must be 30 A rating and **not** 15 A. If a number of sockets are to be added then it may prove more economical to run a new circuit.

To run a new power circuit is a relatively simple process. First a new consumer unit large enough to run the circuit from must be fitted or, alternatively, a consumer unit fitted large enough to accommodate all the original points as well. Then, equipped with a bolster chisel, flat chisel and lump hammer, cut into the brickwork to a depth of 50 mm where the sockets are to be fitted. At this point it is perhaps advisable to cut holes large enough to fit double sockets rather than single ones. The hole in this case needs to be 175 mm wide by 50 mm high.

If the floor is made of wood the socket holes should be cut near to the skirting board, and in this way only a small amount of plaster will need to be removed between the hole and the top of the skirting board. If a very thin chisel is used it is possible to force this down behind the skirting board, thus removing the plaster. Considerable force will then be needed to force the chisel through the piece of floorboard. Once this has been completed the cable can be fed from under the floorboards, behind the skirting board and up to the socket hole. This makes for a far neater job than if the skirting board were to be removed or, indeed, if the cable was fed over the skirting board and into the socket. This latter method not only looks ugly and unfinished, but may also be dangerous as the socket can easily be caught by persons walking past or by a vacuum cleaner being pushed hard against the cable which could cut or chafe the insulation.

If the downstairs floors are solid then, unfortunately, the cable must be fed between the joist space and ultimatedly down the wall, either on the surface or chased into the wall, and protected by capping. If this is the case it would be beneficial to mount the sockets level with the appliances to be fitted into them, and an ideal height seems to be between 300 and 600 mm from the floor.

The socket boxes can be fitted either by pushing out the circular disc which is meant for the cable entry and then bedded in cement (this will hold them quite firmly) or, alternatively, they can be secured by a wall plug and screw.

If the cement method is used, the boxes will need to be left for a couple of days to allow the cement to set hard. If difficulty is encountered when bedding the cable under the floorboards due to it getting caught in the joist stringers (small pieces of wood used to keep the joists vertical and parallel) a piece of oval-shaped capping which is extremely flexible can be fed from one end of it. If the 2.5 mm^2 cable is folded into two it can be forced down the end of the capping; on withdrawing the capping the cable will also come with it. The capping is made in 3-metre lengths, therefore this method only reads good for runs of up to 3 metres, runs of a greater length will need another floorboard removing to provide a halfway point.

If the new power points are to be added to an existing ring main then all the socket faces can be connected. A good guide to the amount of cable that should be left for connection is to place the socket face under the box and at 90° to it (i.e. sticking out from the wall), the cable should be 15 mm longer than the end of the socket. The outer sheathing then has to be stripped back exposing the live and neutral conductors, both insulated, and the non-insulated earth wire. Following this strip back the insulation on the live and neutral conductors for approximately 15 mm (at this point care must be taken not to cut into the actual conductors), these can then be connected into the live and neutral terminals in the back of the socket. The earth wire which is insulated must be sleeved with PVC green sleeving to prevent it from accidentally touching one of the other two terminals which would result in a short circuit, and then connected to the earth terminal. When this has been done the socket face can be screwed on to the box, but care must be exercised to feed the cable gently into the box making sure not to trap any between the socket face and the edge of the box. As the socket face is slightly larger than the box it is a good idea to cut a hole carefully into the brickwork; once the socket has been fitted it will hide the coarse edge of the brickwork.

Once all the sockets have been connected the power should be switched off and the two pieces of cable forming the start and finish of the new circuit should be connected into the two 30 A joint boxes. Following this the power can again be switched on and the new sockets will be an integral part of the ring.

If a new ring is to be fitted then the method would be the same as just mentioned, except that the two ends of

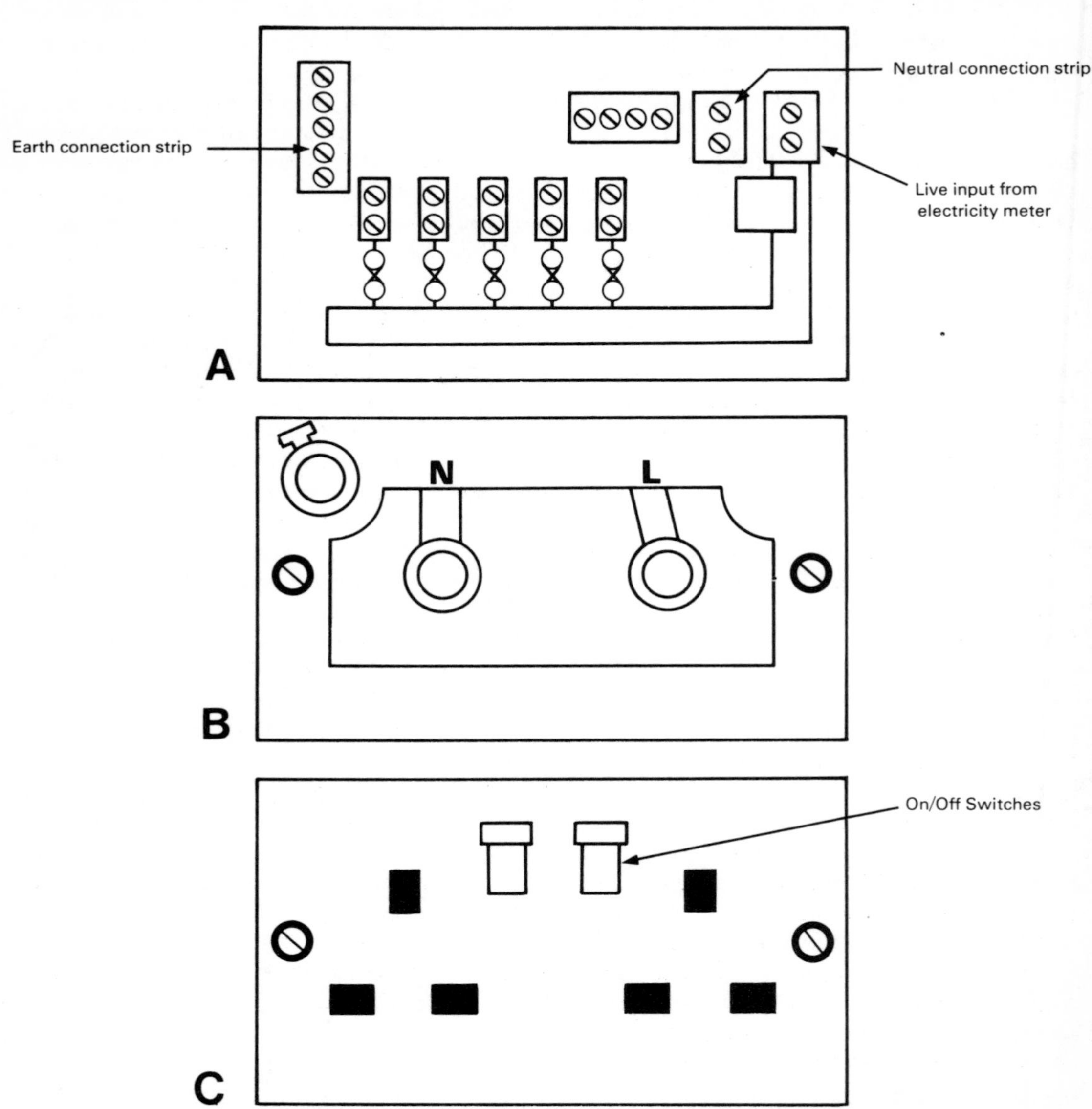

(A) Consumer connecting plates; (B) Rear of socket showing terminal connections; (C) Front of 13 amp double switched socket

the cable would be fed into the consumer unit and ultimately a 30 A fuse.

If a garage is built on to the house a socket can be fitted into the garage as part of the ring but, as mentioned previously in the lighting section, if the garage is some distance away from the house then a special circuit must be fitted.

Wiring a Garage for Lighting and Power

As with all additions to the existing wiring circuit, a new consumer unit must be fitted to accommodate the fuses needed to supply the garage. The cable or cables, whichever the case may be, can be either fed underground or supported on insulated pillars and strung from the side wall of the house to the garage.

To run the cable underground one must dig a trench of at least 2 spade depths deep from the point of emergence from the house to the garage. 2.5 mm² cable can be used but this will need to be protected from both the weather and any insects or even rodents. To protect the cable either galvanised steel conduit piping or the more modern type of conduit which is made of plastic can be used. Both types consist of a tube made of metal or plastic and can be obtained in sizes ranging from 15 mm to 50 mm diameter.

The steel type has screw threads cut on either end and a series of connectors which are steel tubes 30 mm long with a thread cut on the inside used for joining two lengths of pipe together. Square bends can also be purchased in order to allow the pipe to be run in any desired direction. The steel pipe can be bent but this requires a special pipe bender and a large capital outlay would be required to purchase one. Hiring this bender could prove as expensive as engaging an electrician to do the work for you. It is, therefore, far simpler and more straightforward to use galvanised piping. Plastic tubing utilises the same principles as the steel but is cheaper and probably easier to work with. Although not to be encouraged, it can be bent, if it is slightly heated and bent gradually and with extreme caution. If, however, the plastic is heated too much it will be found that on trying to bend it it will collapse. The best method of heating the pipe is to hold it over a gas ring or a small blowlamp.

An alternative to the method of threading normal 2.5 mm² cable through the conduit is to use a special rodent-proof underground cable. The outer sheathing is made of special heavy gauge PVC to help protect it from the minerals in the earth and from rodents and pests. The heavy gauge PVC does offer some resistance if at any time someone should inadvertently dig deep and come into contact with the cable. An idea to mark the presence of such a cable is to sprinkle a coating of fine gravel over the cable before replacing the soil. This will give a visual check of its presence.

Another method of feeding a cable is to use mineral-insulated PVC coated cable; this is an expensive type of cable and requires specialist tools to seal the termination ends. This cable basically consists of three single strands of copper wire encapsulated in a white insulating powder which are all encapsulated in a circular copper sheath. The copper sheath is then coated in PVC for added protection. As stated, this cable is expensive and does require specialist tools in order to fit it, but I mention it

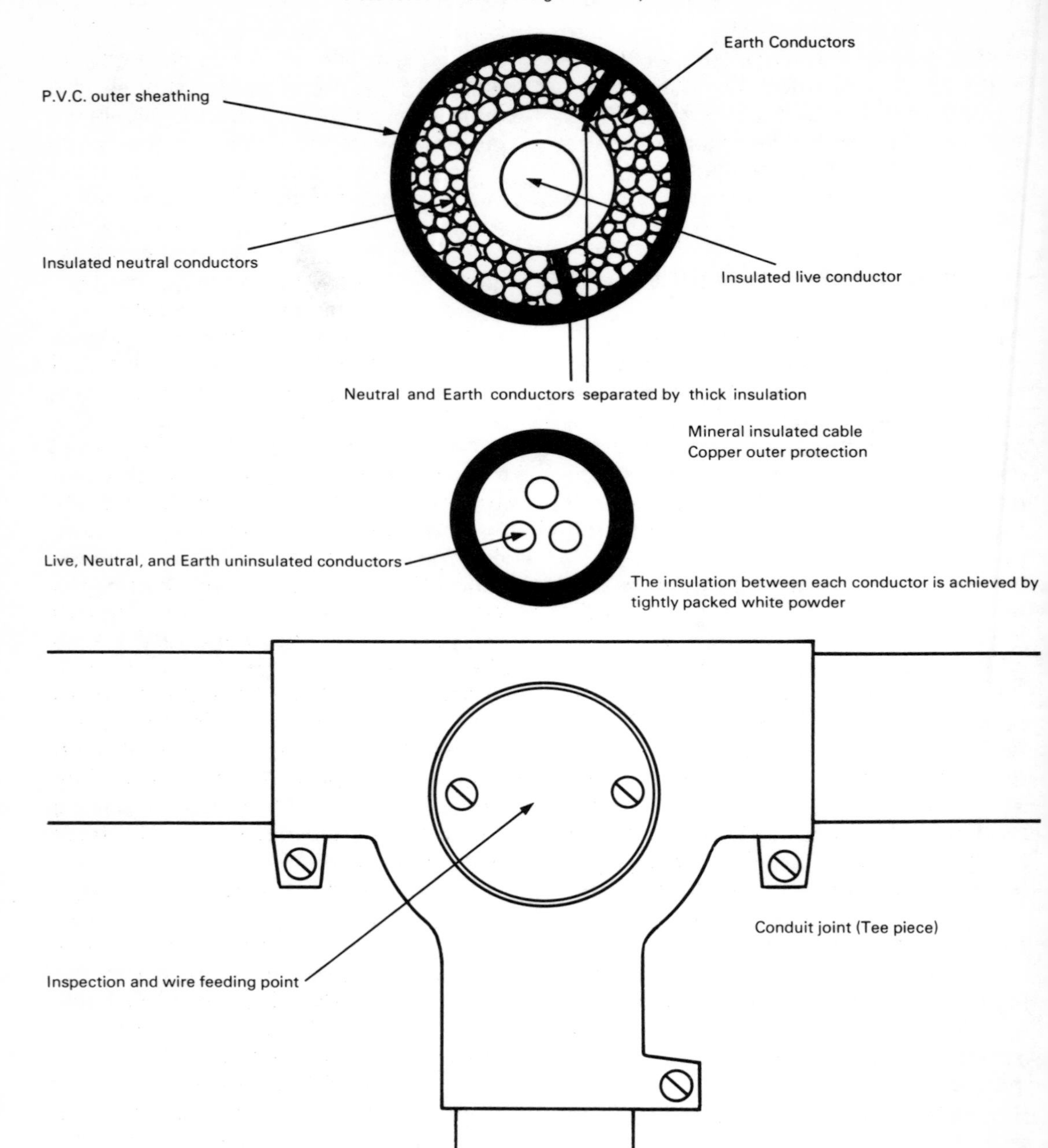

Cable and conduit fitting

as an alternative should anyone have easy access to such tools.

If the cable is run overhead from the house wall to the garage this cable, normally 2.5 mm² PVC, needs to be supported by a piece of heavy gauge wire, similar to the old type of wire clothes line. At the emergence of the cable from the house an insulated support must be mounted on the wall and the cable should be securely wrapped around the support. The supporting cable should be fitted to the wall and run to the garage wall and supported there; the cable should then be taped to this at not greater distances than 450 mm. Finally the cable is wrapped round the insulator mounted on the garage wall and fed through the wall into the garage.

Of the two types of cable feeding, the underground method, although involving more work, is to be preferred as it makes for a much neater appearance when completed.

Having now discussed the methods of feeding the cable to the garage it must be decided how many power points and/or lighting points are needed. If lighting and power are required in the garage then ideally two cables will need to be run to the garage, 1.5 mm² for lighting and 2.5 mm² for power. If, however, only one socket and one light is required then a simple circuit can be used. For this a piece of cable should be run from the power main (see the previous section on power circuits—one spur with two points connected to it). One single piece of cable can be run to the garage and into the power socket. From the socket a piece of 2.5 mm² must be run to a cord outlet spur box (see section on sockets) and from the cord outlet spur box a piece of 1.5 mm² can be fed to the light rose required—**make sure the fuse in the cord outlet spur box is 5 A.**

Once the electricity has been connected to the light rose it only requires a switch and switch wire to be fitted to the rose to complete the circuit. It must be emphasised that under no circumstances can this circuit be added on to as it conforms only to the minimum of requirements and, although basically acceptable, it is not the most preferred system.

It is possible that the garage will need two or three lights plus two or three sockets, in which case a separate ring main for both lighting and the power circuit will be required. Two spare fuse spaces should be available in the consumer unit; if not then a new two fuse unit should be installed to enable two separate ring circuits to be run to the garage and back again. Remember, the lighting circuit should have no greater than eight lighting points on that circuit, but the ring power main can have as many sockets as one wishes providing that the floor area does not total more than 100 square metres (1000 square feet). This is the best method in which to feed a garage with both lighting and power sockets, as the circuits can be extended to include a greenhouse, if desired, by cutting into the ring and using joint boxes (as described on pages 46–8).

Earthing

Earthing is probably the most important part of any electrical installation. The main purpose of the earth is to prevent overloading in a circuit and help eliminate the possibility of a person being electrocuted. The current will, if allowed, find the easiest path to earth in the event of a fault occurring in the circuit.

The continual earth wire which connects all the sockets or light fittings in one continuous loop is ultimately fed into the consumer unit and from there to the earth tag on the Electricity Board's armour sheathing on the input cable. The resistance of this wire is something in the order of less than 1.5 ohms. Maintaining the resistance at this low level will ensure that any stray current will flow along this wire to earth. The resistance of a human body can vary from between 10 kΩ (10,000 ohms) to 100 kΩ (100,000 ohms). It can therefore be seen that the presence of the earth wire, which is something in the region of 20,000 times lower resistance than in a human, does ensure that a shock will not affect a human being. Should no earth wire be fitted in the circuit, then the only path available for the stray current would be through the human; this would at the least result in a very nasty shock and at worst in fatality.

Earthing regulations up until 1973 were such that the lighting and power circuits had to have a continually uninterrupted earth loop on all circuits. The lighting circuit earth wire, which is fitted into both the light rose and the light switch, is mainly for the prevention of an accident occurring should a steel light fitting be used. Should a short occur in a steel light fitting, then the omission of an earth wire would mean that when one went to change the bulb the circuit would be taken down to earth and a nasty shock, which again could be fatal, would result.

The earth loop fitted to the power main has provision made for its connection into the circuit by producing the 13 A socket faces with three terminals, one for each—live, neutral and earth. Lighting circuits are somewhat different inasmuch as the lighting switches have no facility for the earth conductor to be actually connected to the switch. However, there is a terminal fitted in the switch box which is mounted on the wall. Ceiling light roses have an earth terminal mounted in the back plate of the rose which is separate from the other three sets of terminals used for connecting the live and neutral conductors; this layout does provide a continuous earth loop round the lighting circuits. Both earth terminations are ultimately connected on to a common earth terminal in the consumer unit.

The connection of the earth terminals from the consumer unit to the earth tag or stake can be carried out in three or four different ways, and these are best dealt with separately in detail, as follows.

1. Using the Electricity Board's Armour Outer Sheathing

If the property was built approximately before 1973 the earth connections must be made in the following manner. First of all the earth on both the lighting and power circuits must have a continuous earth wire which is fed into the consumer unit. All the earth wires are connected on to the same common terminal and from this common terminal a piece of 6 mm^2 single cable—colour coded yellow and green—must be fed to the input side of the gas meter should one be fitted, and another to the input side of the mains water pipe. If, however, the input water main is made from alkathene pipe (plastic) then an earth connection to it would not provide a good conductive connection as the piping material is an insulator; the earth wire is therefore connected on to the copper pipe just before the main input stopcock. This will provide a good conductive connection and the water inside the copper pipe will then conduct the current along the alkathene pipe and ultimately into the main water pipe situated somewhere in the road. This water pipe is made of cast iron, again another good conductor of electricity, and will therefore take the stray current directly to earth. A third wire, again 6 mm^2, must be fed to the earth tag

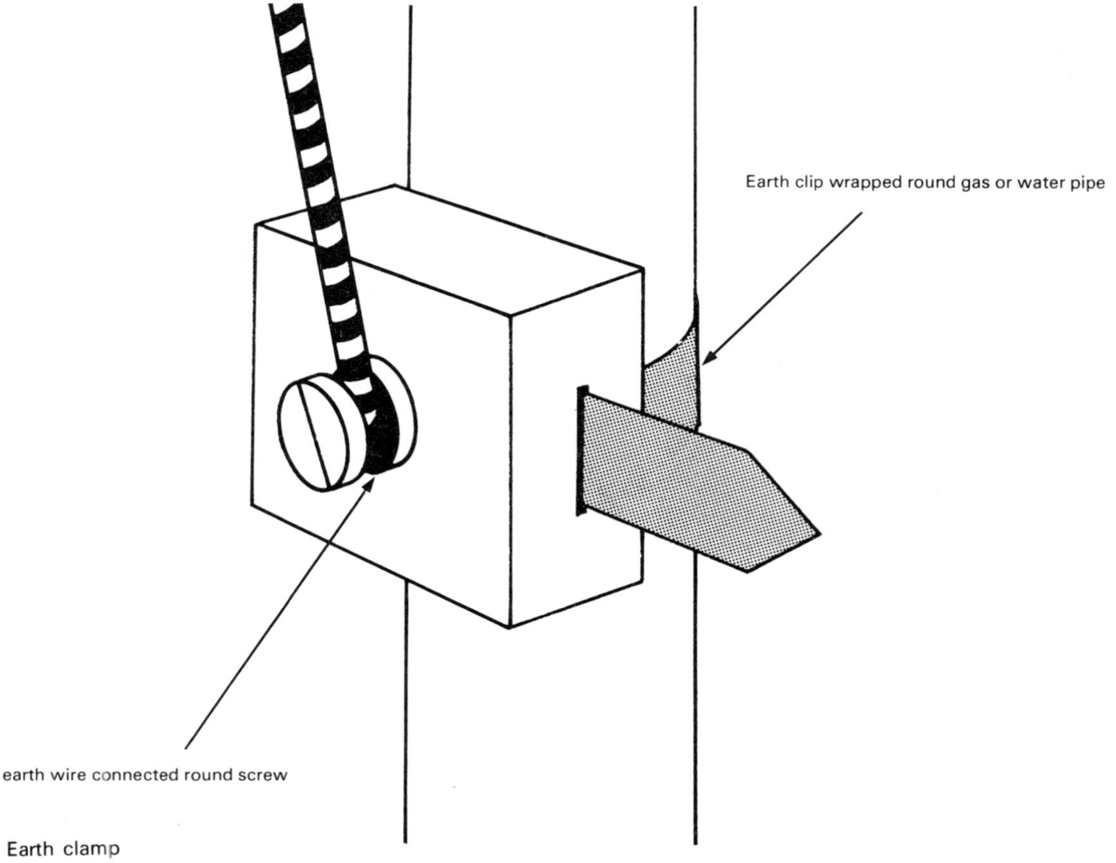

Earth clamp

soldered on to the Electricity Board's armour-protected sheathing. By connecting the earth leads to three sources, which ultimately go to earth, there is multiple protection against the risk of an overload. The clamps used to connect the earth wires to the various pipes must be of a recognisable standard, and any electrical factors will assist in the purchase of these.

2. Multiple Protective Earth

Houses which were built in 1973 and after will have a protective multiple earth system fitted. This is similar to the previous method inasmuch as the gas and water pipes must be earthed, but in addition to this the sink top which is made of either stainless steel or enamelled steel must also be earthed. Most modern sink tops are fitted with a short strip of steel which has a number of holes drilled in it to which the earth wire can be bolted. As the water stopcock is nearly always situated under the sink, bonding the sink top does not pose any considerable difficulty or inconvenience. If, however, the stopcock is fitted in, for example, a cloakroom, then a separate wire must be fed to the sink top. The old idea of bonding all the taps and wastes to the bath leg can be omitted as this no longer constitutes a requirement, since most modern properties are now fitted with plastic waste pipes.

The Electricity Board, instead of providing an earth tag on their armour cable, have produced a new type of main fuse. In old properties the live and neutral wires were fed into the main fuse which was of either 30 A or 60 A (nowadays a 100 A fuse is used) and then on to the meter, the live conductor having the fuse in the circuit and the neutral being fed straight through. Under this old system it meant that the voltage measured between the live and neutral was 240 V, but in some instances it was possible to produce a voltage between neutral and earth. This was probably only something in the region of 3 or 4 V but to the amateur radio enthusiast or the keen student studying electrical engineering, it meant that they had a free supply of electricity which was useful for running an internal telephone system or intercom and, of course, as the voltage was produced between neutral and earth there was no means of it being monitored and the Electricity Board was therefore being cheated. To prevent this from happening, and in the interests of safety, the Electricity Board have now fitted a main fuse which has the facility for the live to be connected to the main fuse while the neutral and earth terminals are internally linked, which ultimately puts the neutral and earth at the same potential. This type of earthing is known as multiple protective earthing and most electricity authorities will provide a pamphlet dealing solely with this type of earthing.

3. Earth Leakage Trips

If the main input cable is fed into the house by overhead means, which is still common practice in some rural areas, then there is no provision for an earth, as an earth fitted from a pylon would not be suitable to take any stray current down to earth and ultimately to safety.

In the case of overhead power systems, the normal domestic wiring has its continual earth run uninterrupted round the circuit and is fed into the consumer unit in the normal manner. From the consumer unit there must be an earth wire connected to both the gas and water pipes and another earth wire must be fed to an earth leakage trip.

The earth leakage trip is an appliance which will trip out and render the circuit safe if an overload or current is present in the earth wire. These trips have four main terminals and into these the live and neutrals are fed from the Electricity Board's main fuse and then out of the earth leakage trip to the main consumer unit. There are also two terminals marked 'F' and 'E'; in the case of overhead wiring the 'E' terminal is used. An earth wire is run from the consumer unit to terminal 'E' and subsequently to a copper stake which must be hammered into the ground, a complete circuit then being formed via the earth leakage trip to earth. A small current flowing along the earth wire does suggest a fault in the circuit and therefore the trip will trip out, cutting the live and neutral to the consumer unit. Sometimes, however, a minor fault will occur in an appliance such as an electric kettle and the trip will trip out indicating a faulty circuit when, in fact, it is the appliance at fault. When an electric element becomes old the insulation in the element sometimes loses some of its insulation properties and a small current flow is then set up between neutral and earth resulting in the trip tripping out. If this is found to be the case then the only thing to do is to keep re-setting the trip until the appliance becomes so bad that it persistently trips the circuit. When this occurs then it will be necessary to replace the element.

Earth leakage trips have been known to trip out due to vibration, and people living at the side of a busy trunk road have found this to be a positive nuisance.

Earth leakage trips can also be used in houses where the cold water system has been installed in plastic pipes, thus allowing no facility for an earth wire to be connected on to it. If, however, the mains input cable is fed from overhead and the water pipe is plastic then both the 'E' and 'F' terminals of the earth trip need to be used. Two wires are fitted from the consumer unit, one to terminal 'E' and one to terminal 'F'. These two wires are fitted to two copper stakes which must be placed at least two metres apart thus providing two separate earth paths for any stray electrical current.

Should the reader be in any doubt as to what system is employed in the property in which he finds himself working, then the earthing regulations can be obtained from any regional office of the Electricity Board. However, these regulations have not been drawn up in order to make life difficult but have been produced with one fact in mind, and that is to provide a safe and efficient multiple path for any stray electrical current to pass directly to earth thus safeguarding anyone from becoming electrocuted. So remember, never flaunt the rules by trying to be clever and leaving out any of the requirements stipulated to effect a safe and reliable earth system.